Self-Sufficiency

Keeping Chickens

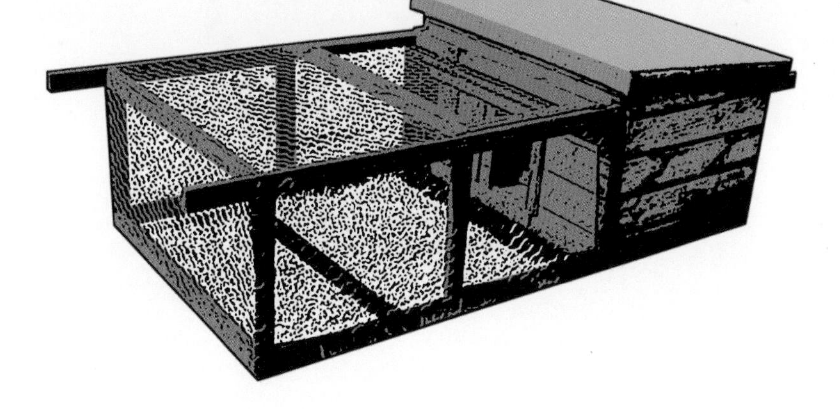

Self-Sufficiency

Mike Hatcher

Skyhorse Publishing

Skyhorse Publishing books may be purchased in bulk at special discounts for sales promotion, corporate gifts, fund-raising, or educational purposes. Special editions can also be created to specifications. For details, contact the Special Sales Department, Skyhorse Publishing, 555 Eighth Avenue, Suite 903, New York, NY 10018 or info@skyhorsepublishing.com.

www.skyhorsepublishing.com

10 9 8 7 6 5 4 3 2

Library of Congress Cataloging-in-Publication Data

Hatcher, Mike.
Keeping chickens : self-sufficiency / Mike Hatcher.
p. cm.
Includes index.
ISBN 978-1-60239-977-8 (hardcover : alk. paper)
1. Chickens. I. Title.
SF487.H336 2010
636.5—dc22
2009046854

Printed in China

CONTENTS

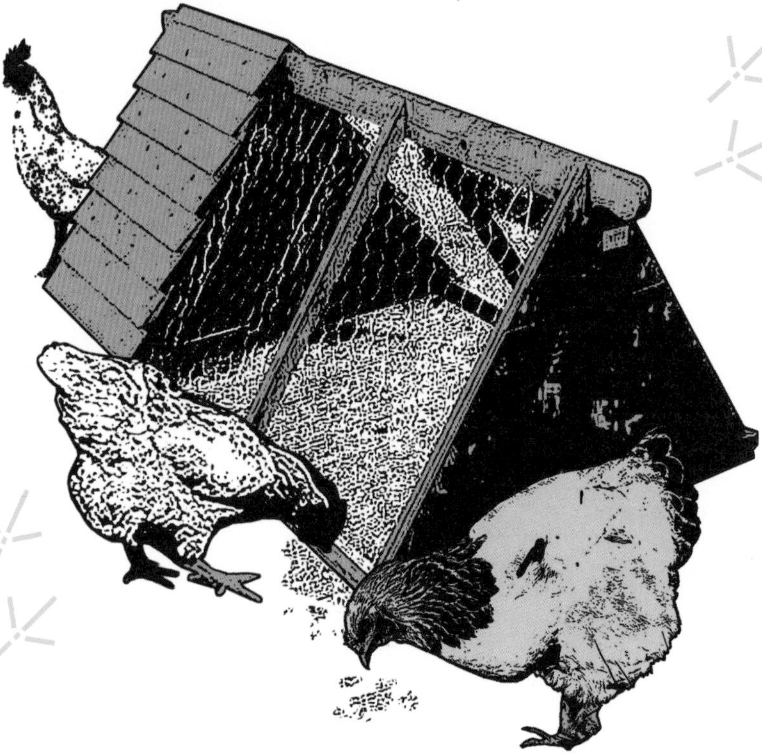

Keeping chickens is very satisfying on many levels and there are numerous benefits to be enjoyed. This book is aimed at those who have contemplated keeping a few hens but have no idea how to get started. Make no mistake: Keeping chickens is not a casual hobby you can pick up and put down when you feel like it! It is a genuine commitment and takes time and dedication.

The basics

The main reason many people wish to start keeping chickens is to enjoy a regular supply of freshly laid eggs. The idea of having your own fresh eggs produced by birds you have fed and nurtured yourself is very appealing. A little hen house and run with five or six hens sounds straightforward enough, but you need to be fully acquainted with the dos and don'ts of keeping chickens to ensure your brood is happy and will lay regularly.

The rewards of keeping chickens

As with all living things, hens will not thrive unless they are cared for. A commitment to the welfare of your birds is essential if you are to find satisfaction in keeping chickens. To be successful, you or someone you trust has to look in on your birds every day. The idea that simply providing hens a house, food, and water, and then collecting eggs at the end of the day, is not what looking after hens is all about. You have to be more interested than that and I hope that after reading on and keeping a few birds yourself, you will find the enjoyment that so many others have experienced.

The reaction of anyone on first seeing a live chicken in close quarters is very interesting; most are amazed at the detailing on the plumage, its patterns and formation, and at the variation of heads, eyes, and legs between breeds. Touching a chicken often produces a reaction of wonderment! Until you see and feel the real thing, a photograph does not tell the whole story. So be ready for it, whether it is a baby chick or a large, fully-grown hen—they have a warmth and texture that is soft and very appealing.

You can purchase birds at any age, from day-old chicks to laying birds. For simplicity's sake it is easier to buy young hens, known as pullets, at point of lay. However, for a bonding and real learning exercise, especially if you have young children, get baby chicks. You and younger members of your family will enjoy handling and tending them, and watching them grow can be a real educational process! Not only will it teach you and your children the essentials of life but also the practice of caring for and loving all living creatures.

Where you live

Where you live can play a major part in how successful you are with your keeping chickens. Unless local bylaws prohibit the keeping of pets on your property, then you should be free to keep hens. If you have close neighbors, it may be advantageous to inform them of your plans. If they are persuaded it is a good idea, then try to get them involved. Invite them to stop by when your birds arrive and promise them that if you have any spare eggs they will be the beneficiaries!

You will need sufficient space for your birds' welfare. Many people are motivated by the idea of taking hens out of the intensive conditions seen in battery cages. You do not need a grass surface for your birds although most will opt for one. Remember, chickens will not only eat the grass by pecking it but will also wear it out by scratching to find grubs. If you decide to let your hens roam in the yard be ready for them to disturb your small plants or peck at any vegetables you might be growing. Chickens can be left to roam freely on a patch where vegetables have already been harvested. Their constant pecking and scratching will keep the area clear of weeds and harmful insects until the next crop is put in; the ground will also be naturally manured. A light soil such as a sandy loam will be better in a wet time than a heavy clay that tends to get very muddy.

Types of birds

Before you decide which birds to go for, you have to be aware of the many options available to you. At supermarkets you will notice that you can buy eggs of various sizes. This is usually governed in the commercial world by the age of the bird: When young hens start to lay, they produce small eggs. No egg producer breeds a hen to lay small eggs when it will still consume the same amount of food as a hen that lays large eggs. However, bantam chickens, which, when fully grown, are about a third of the size of large hens, will lay eggs that are about half the size of those of large hens.

In the large hen category, many hybrids are bred specifically to lay plenty of eggs on a minimal amount of food. You may decide to go for one of the pure breeds that come in a myriad of colors and shapes, and have added visual impact over the hybrids. They do not, however, lay as many eggs as hybrids.

Different breeds produce eggs in a great variety of colors: The shells range from a deep chocolate through to a pearly white, blue, and even green! The contents are the same though. These colors are also available in the bantam varieties. The advantages of breeding bantams are that they require less space and eat less food than hybrids. Also, because they are smaller than hybrids, bantams can be easier for children to handle.

Choosing your birds

You must now consider how many birds you would like to keep. Estimate how many eggs you buy at the supermarket per week for you and your family, then add a few more, as you will likely enjoy your own eggs more than store-bought ones, as well as the convenience of having them on hand. Your average chicken will, if looked after properly, lay, on average, five to six eggs per week. Too many eggs can become an inconvenience as your family and friends will not always want those extra eggs! Chickens are not solitary creatures, so it's a good idea to keep at least two chickens, as a chicken on its own will feel lonely.

Point-of-lay pullets

You may decide to take the easy option of buying fully-grown young hens, which in the poultry world are known as "point-of-lay pullets." The advantage of buying point-of-lay hens is that they accept a change in environment more readily and suffer fewer setbacks than hens that have already started laying, who may stop laying for a few weeks as a result of this disruption.

So where can point-of-lay birds be purchased? The Internet is always a valuable source of information, but sellers advertising in specialist poultry magazines tend to be more reliable. As there is no restricted breeding season for chickens (they lay eggs that will hatch into chicks all year round), birds are always in plentiful supply. If ordering your birds, however, it is a good idea to check availability in advance, as it takes about four months to rear point-of-lay pullets and they can sometimes be sold out. If your source is quite local, then it is advisable to collect them yourself so you can see what you are getting. If you see a contented bunch of chickens running around looking happy and healthy, then so much the better. On the other hand, if they look hunched up and inactive, you should exercise caution. Do not expect to be able to handpick the birds, especially if you are buying direct from the breeder. You must be aware that they will wish to keep the best ones for themselves to retain the quality of their stock.

Day-old chicks

Another option is buying day-old chicks, in other words babies that have only just emerged from the shell. This can be both a challenging and rewarding experience. The challenge comes with finding a good, reliable source of keeping the chicks warm while transporting them to their new home, and the fascinating period of rearing them to point of lay. Unless you find a source where you can buy chicks ready sexed you should order at least three times as many chicks as you intend keeping as laying hens. On average half the chicks will be males, so that out of fifteen chicks you could have eight males and out of the remaining seven you may lose one or two through accidents or other unexpected problems. Only if you are buying from a large commercial hatchery can the sex of your baby chicks be guaranteed. A small hobbyist poultry keeper, on the other hand, will not be able to identify the sex of the young chicks and they will be sold unsexed.

If you do end up with more males than you need, you will not be able to rear them all and keep them among the females, as they will fight each other to mate with the females. You will either have to harden your heart and rear them up until they are sufficient size to be killed for the table or sell them to someone else who is interested in doing just that.

For the first forty-eight hours of their life the chicks can live on the yolk sac from the egg that is absorbed into the body at the end of the incubation cycle, so do not worry if their appetite is small at the start.

Cockerels

Hens will lay eggs whether or not a rooster is present. Roosters are only needed for an egg to be fertilized (so that it can become a chick). Unfertilized eggs can never hatch chicks, and are simply for eating. Roosters crow and can be a nuisance, especially early in the morning. In most areas, even in the countryside, the crowing of a rooster is no longer accepted as a part of country life. But if you don't have any neighbors nearby and are happy to keep a rooster, you will find, like traffic noise, that you soon get used to its crowing.

Getting set up

Before you bring your hens home, you need to be fully equipped to receive them. You will need:

- A **hen house**. This may come in sections and have to be assembled.
- A good **tool kit** comprising a claw hammer, screwdrivers, an adjustable wrench, a pair of pliers, wire cutters, a saw, and a good supply of basic nails and screws.
- An enclosed cage or **run** to prevent your stock from roaming too far. These are often attached to the hen house. If building your own run you will need the appropriate wood and chicken wire.
- **Nest boxes**. Hens should preferably lay their eggs in specially provided nest boxes, which should be kept in the hen house. This is partly so that you know exactly where to find the eggs rather than having to spend time hunting for them, and partly to keep them protected from predators.
- **Perches** for the birds to perch on at night.
- **Litter** for the floor of the house—wood shavings are acceptable and are usually readily available. These can also be used as padding for the base of the nest boxes.
- A **drinker** to hold water. These are usually of the water fountain design and should have a lip around the base from which the birds drink. If you do not have a tap nearby, then you will need a **watering can** or large **bucket** to carry water to fill the water containers.
- A **food trough** or **hopper** is a good idea, as unless you are feeding grain, any other food will get mixed into the litter. Food containers should be kept in the chicken coop, sheltered from the weather and from wild birds.

- **Feed**. Before you collect or have your hens delivered you should source the food your birds will be eating and have some in stock. Try to find out what the previous owner was feeding them; you do not need the same brand of feed but it is best to use a similar type.

- A **small grit container** to hold mixed grit, which the birds can pick through; hens need a supply of soluble and insoluble grit to help them produce the calcium needed for egg shells and also to grind the food when it reaches the gizzard in their digestive system.

- A **bucket** and **scoop** for transferring feed from the feed bag.

- A **shovel** for cleaning out the hen house and run: a small one with a short handle to clean out a small house, or a large one to clean out a larger house. Spades are not as effective in picking up loose litter.

- A **wheelbarrow** for transporting waste from the hen house and run.

- A **dustpan** and **brush** for smaller tidying-up jobs.

- A **flashlight** for early-morning or late-evening visits to your birds when the days are short.

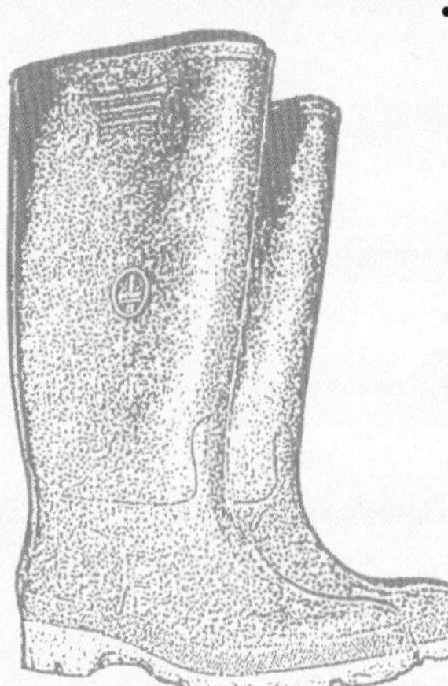

- A **pair of rubber boots** with good grippy soles are essential, especially in the winter, as are waterproof pants, jacket, hat, and gloves. It is best not to wear brightly colored clothing as this may frighten your birds.

We shall look at these requirements in detail in the following pages.

Housing requirements

No matter how many birds you intend to keep you will need some form of housing to keep your birds safe, dry, and warm. Though the thought of birds roaming around freely at all times sounds idyllic, the reality is very different. Both in cities and in the countryside there are a number of predators around to which your birds will be vulnerable, and against which you must provide suitable protection.

Additionally, birds will need to be sheltered from bad weather; although birds regularly anoint their feathers so that they are water-resistant, too much exposure to the elements will reduce the protective quality of the feathers. Also, if a bird is battling with the elements it will not have enough energy left to produce eggs. Once you have taught your birds that they should return to their house to sleep and lay eggs, you can relax in the knowledge that they are safe and sheltered.

Runs

In addition to a house, you will also need a cage of sorts, to prevent your stock from roaming too far. In the poultry world these cages are known as runs. The most convenient sort for a small number of birds (up to six) are those that are attached to houses. These units are usually light enough to be moved to prevent the ground from wear and tear and being fouled with bird droppings. If you do move your run and house regularly you will find that the grass will grow stronger and of a deeper green because of the nitrogen in the droppings. The scratching of your hens will also rake out moss that can often choke grass. If you prefer a static run (which is often attached to the house to form one unit) you will need to fence in a larger space of around 4 ft. 11 in.–6 ft. 7 in. with wire netting, depending on the type of birds you decide to keep. You will need good strong posts and a gate wide enough to get through with a wheelbarrow, which you will need for cleaning out the hen house and run. Electric fencing is an alternative but is not always user friendly, especially in small spaces.

If you own or have access to a large enclosed field or paddock, then you may decide to have just a house without a run. You must, however, take extra precautions, as foxes and other determined predators may find a way of entering the enclosure, putting your birds at risk.

Converting existing outbuildings

Hen houses can be expensive, so before purchasing housing for your hens it is wise to see if there are any outbuildings or sheds on your property that could be adapted for this purpose. If you are fortunate and live in the countryside you may have a redundant brick or concrete-block building that could be adapted. It would need to be fairly sound with a waterproof roof and solid walls. The floor could be concrete or even soil, but should be level for ease of cleaning. A window is an advantage, although if you have a door with the top half covered in wire netting it will provide enough natural light and ventilation.

It is most important that any type of building you use has good air circulation to prevent respiratory problems, as most poultry afflictions revolve around the respiratory system (see Diseases, pages 62–4). Also, a house that is dark and closed up will not be the most pleasant or salubrious place for you or your birds to spend time in. If you have a permanent building with electric lighting, then so much the better, as the additional light can help keep up egg production during the dark winter months. Light stimulates the pituitary gland at the base of the brain that helps to promote laying. Hens are at their most productive when they receive an average fourteen hours of light a day.

If you do not have a brick building, then a similar garden shed could be adapted. You will need to drill a number of holes under the roof-line of the shed for ventilation purposes to remove stale, warm air. For additional ventilation you could also make what is called a chicken hatchway. Serving a similar purpose to a cat flap, this is a hole in the wall or door large enough for a hen to push through. You can then fit a sliding or hinged cover over it. If your shed has a glass window, it is worth making a wooden frame of the same size, covering it with wire netting, and fitting it to the inside of the window. This is so that if a hen flies against the window it won't break the glass and harm itself.

Large wooden or plastic boxes, at least 18 in. square in size, can be used for egg-laying purposes. You can easily make your own wooden boxes from plywood; plastic ones can be purchased from retail outlets.

Fix a number of perches, made from lengths of timber that are 2 in. thick in cross-section, positioned about 3 ft. 3 in. above the floor along one or more of the inside walls of the house (depending on how many birds you have), for the birds to perch on at night. Each bird will need approximately 12 in. in length of perch. Leave sufficient space between the wall and the perch (about 8 in.) for the birds to comfortably settle on at night. Make sure you do not place the nest boxes underneath the perches as the birds' droppings will soil the eggs.

This may seem like a lot of work if you are not handy with tools, so the idea of purchasing a ready-made unit might be appealing.

Ready-made hen houses

No matter how beautiful a house may look in a catalog, try to see it before buying it and ask yourself the following questions: Can you comfortably get into it to clean it out without dismantling it? Is it so heavy it needs a team of people to move it? Or is it so light and fragile that a strong wind could blow it over and break it?

There are many types of hen houses on the market. You can have a small house and run as a single unit or a separate house and run, which, if you are doing the chores on your own, might be easier to move. Many houses do not have separate nest boxes as they are multipurpose and can also be used for rearing young birds.

Most hen houses are made of wood, usually sawn boards on a timber frame. These boards are usually thinner at one edge and are lapped over the lower board where they join. These can be fine if the timber has been seasoned but if unseasoned the boards are prone to warping and can therefore leave gaps that a predator can open. Some houses are made of plywood, which has its advantages, but it must be exterior ply and weatherproofed. The most expensive houses use smooth planed wood boards that are jointed together. These have the advantage of being draught-proof and are easier to clean than the sawn boards.

Apex houses

Houses come in many shapes and one of the most stable and strong is the apex type. This design is well proven and is quite often cheaper than the square designs. It gives a good floor area and although it may look cramped at the edges the hens can still reach to the side to peck the grass. The house usually has air vents at the apex covered with some form of wire and a solid floor, but of course the run area has no floor so the birds can peck the grass.

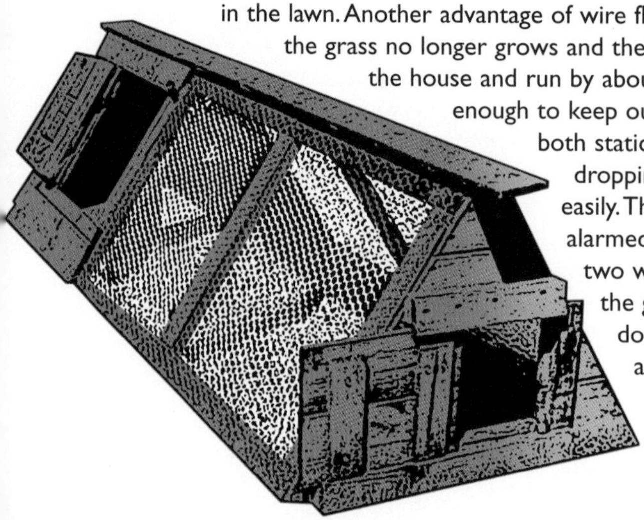

A useful addition to these houses is a wire floor to the run. This protects against predators digging under and will stop the chickens digging a hole in the lawn. Another advantage of wire floors is that in the winter, when the grass no longer grows and the ground is wet, you can raise both the house and run by about 12 in. on bricks or bearers, just enough to keep out mud and droppings, and leave both static. This way the birds stay clean and droppings can be cleared away more easily. The birds do not seem to be alarmed by this raising, and after a day or two will quite readily adapt to being off the ground. If, on the other hand, you do not have a wire floor to the run and choose the static option, then a layer of wood chips can be laid down in the run during the winter months and the birds will scratch among the chips, incorporating their droppings. Depending on the depth of the layer of wood chips you lay down, they should be cleared away either every day (if the layer is thin) or up to once a month (if the layer is thicker). This makes good compost for your garden.

Rectangular houses

The next type of house and run is rectangular in shape with a flat-topped run and sloping-roof house. Many runs of this design have a lift-off top to gain easy access to the birds and equipment. This type of house has a hinged or sliding roof. Although it may not be particularly convenient, a sliding roof has the advantage of offering the birds a smaller space to escape than the hinged version, which, because of its design, lifts open completely, giving the birds an opportunity to escape more easily. With these houses and most of the flat-roofed versions a protective cover is placed on the roof to make it waterproof. Many wooden roofs are covered with bitumen felt to prevent water penetration. These roofs should have wooden battens at the edges to stop the wind getting under and lifting the felt off the wood and exposing it to the elements.

Most popular these days is a corrugated bitumen product that has various trade names. It has the advantage over the old corrugated galvanized tin of being easier to cut to size and also sharp edges are not so dangerous. It can, however, be damaged more easily. Corrugated tin has been known to last for up to half a century if given a protective coating of tar or similar preservatives every four or five years. It is wise to check that the nails that fasten the roof covering are all in place; it only takes one loose nail for the roof to lift off. Whichever roof you have, make sure that when you open it, especially to clean out the inside, you do not strain the fixings. It is not so much an issue with sliding roofs but if strain is put on a hinged roof, it can crack the timber it is screwed into, weakening the whole structure.

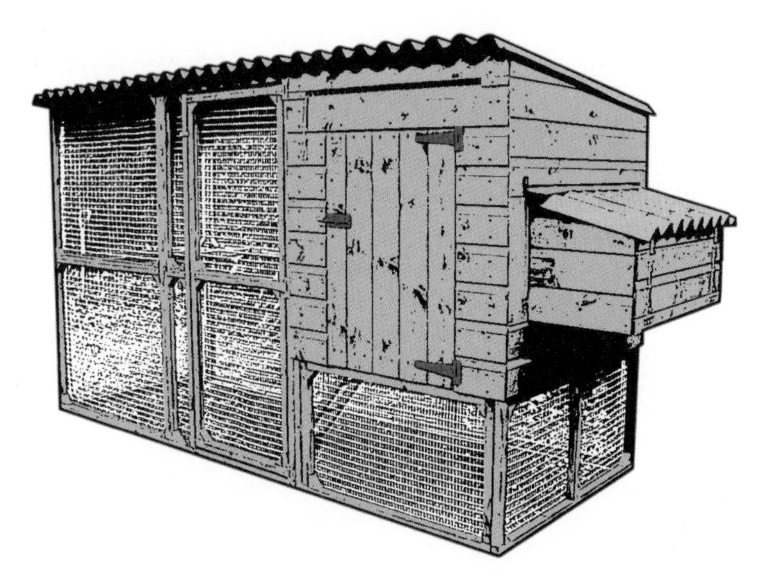

There are other variations on the rectangular structure where the house is contained within the run. Many of these units have the house raised up above the run, which makes it more convenient to clean, feed the birds, and collect the eggs as you will not have to bend down. It also provides a covered area under the house where the birds can shelter during the day. Usually these houses have a long board with rungs across it providing the birds access into and out of the house, but until your birds are used to making their own way up to the house you may have to pick them up for a night or two and put them in the house. Because of their design (heavy timber house at the top and open lightweight wire at the base), these units are top heavy and the wind can get under and turn them over, so either place them in a sheltered position or be ready to use pegs and cords to stabilize them. If your house has an external nest box attached to the house, it is best to have it fitted with a safety catch to prevent predators lifting the lid and stealing the contents.

Stationary houses

If you opt for a stationary house, then the positioning of this needs to be carefully considered. You can make the house part of the outer perimeter fence of the area that the birds will live in. This means that if you have the door on the outside for access to the inside of the house all feeding and cleaning out can be done from outside the wire. You can also have the chicken hatchway where the hens go in and out positioned in such a way that you can reach over the boundary fence and lift and lower it from outside. Some poultry keepers have a long cord that stretches all the way to their own house so they can lift the chicken hatchway without needing to go too far out into their garden. Being stationary, these houses are usually bigger and your birds will have more space in which to shelter in bad weather. In the majority of cases these houses will be rectangular with a sloping roof. A simple design is easier to maintain than one with lots of complicated extras. All you need is a door, usually hinged, one or two windows covered with wire netting, and, if possible, a sliding glass or plexiglass sheet to slide over the windows in case of rain or snow. As long as the weather is dry, even if it is cold, it is better to allow as much air circulation as possible.

Positioning the house and run

If you are positioning any type of movable house and run over a grassy area, then you should aim to move the unit at least every three days but ideally every day. Never let the grass get too worn down. Hopefully your unit will have handles, which, if stout enough, will make short work of the job. Moving the unit sideways means that you will only have to move it the width of the house and run until you reach the boundary of the new area. If you have one of the larger units with retractable wheels, then you will be moving it forward or backward onto fresh ground. These are, however, designed to be used on large areas and you would probably be keeping around twenty-five hens to make the cost of such a unit viable.

Raising the house off the ground by about 12 in. with bricks or concrete blocks prevents rats or other vermin from finding shelter. Being raised off the ground also helps to keep the floor dry to prevent it rotting. As the hens (and yourself) will be coming and going around the area, it is recommended that a pathway made of concrete slabs is constructed around the area. This has the advantages of being stable and easy to sweep or hose clean. The area under a raised house, when it is dry, will become the birds' dust bath. This is where they scratch and roll over to help prevent mites living in their feathers.

Litter

Chopped straw, shavings, sawdust, shredded newspaper, and hay can all be used as litter. Litter absorbs droppings and makes the floor softer for birds to stand on. If your run is positioned over grass, there is no need to cover it with litter. The house, however, should be lined with litter at all times, to provide cushioning for your hens and to soak up droppings.

The bigger the house, the deeper the litter you can put in it, and the less often your birds will need to be cleaned out. In the commercial world, where birds are kept intensively on deep litter, they are only cleaned out once in their year-long laying cycle. The rest of the time they scratch the litter, incorporating their droppings into it. For domestic rearing, you should aim to have fairly deep litter, about 6 in. deep, which should be cleaned out when it starts to get compacted.

It is an idea to place a dropping board under the perching area so that when the birds are roosting overnight, all the droppings are collected in one place and can be cleared away easily. If no board is in place it is best to make sure that any accumulation is cleared away to keep the floor dry, the air fresh and healthy, and the birds' feet clean so they will not soil the eggs. While you are doing this also clean around the chicken hatchway area, which also usually gets very soiled.

Outside shelter

When it comes to summertime and hopefully the weather warms up, your birds will need some protection from the sun. They find life much easier in the cold than in the heat. You may question this after seeing your hens lying out in the sun basking on their sides. A hen cannot sweat to give off heat and the only relief besides panting is to move into some shade. The chicken coop may seem like an ideal shelter but on a really hot day the inside can become like a furnace. You may be using a house with a connecting run; if so, it will be useful to place a sheet of corrugated tin or similar roofing over the top of the run while leaving the sides open for air circulation. With large open runs with trees or bushes in them, these should offer adequate shade. If the house is not sufficiently raised off the ground for them to get under it, a sheet of wood or tin leaning against the house will offer a welcome shady area. You should make sure it is fixed so that in case of strong winds it does not get blown over and cause injury. Just remember, a chicken cannot change its clothes like a human to suit the weather, nor can it take a dip in the water like a duck or a goose.

Keeping out pests

The last thing you want is for your eggs to provide tasty snacks for rats, or, even worse, your hens to become dinner for a fox or other predator. If you can go to a little extra expense, the netting you use should have a strong straining wire at the top and a little turned out at the bottom to prevent foxes from climbing a vertical fence or pushing under the bottom. It is essential that you regularly check and maintain all your fencing and enclosures so that they remain predator-proof. You will also always need to shut your birds inside the house every night. Keeping your stores of feed free from pests is covered in the next section.

Chicken feed

Your local pet store will usually stock poultry feed. ("Poultry" is the general term not only for hens but also for ducks, geese and turkeys.) It is much more economical to buy feed in large 44-lb. bags than in smaller amounts. The ingredients will not lose their goodness during the time that say, six birds, take to eat this quantity of food. Chickens can also be fed household scraps. If you are rearing chicks it may be advisable to buy smaller quantities of special feed.

In order to check up on your birds—to make sure they have come out of the house and are eating—it is nearly always best to feed them twice a day, in the morning and the evening, so that if you go to work it can be done early before you leave the house, and later when you return home. If you are feeding mash in a hopper, then you can fill it once a day or once a week, depending on the size of the hopper, and the birds can help themselves whenever they wish.

The most common and convenient feed is made from a specially ground and formulated meal, and comprises a mix of grains, vegetable proteins, and minerals. These days the grains are sourced from suppliers who guarantee that they are GM-free. Over the years, most grains and forms of vegetable proteins have been selected to make better crops in terms of yield and resistance to disease. However, many feel that there should be a check on scientific experimentation and so it is felt that GM foods should not enter the food chain. For your hens to be truly organic, they should not be given GM feed. A number of the more expensive brands of feed have been developed in response to the demand for organically produced foods. If you want organic eggs you must not only keep your birds in the open air eating natural products, such as grass, but your hens' feed also has to be produced on organic farms.

The meal you feed your chickens comes in a number of forms: meal, mash, pellets, crumbs, and grains.

No matter what you feed your birds—and for best results you should stick with specially formulated rations—a certain amount of feed will find its way on to the floor. If the feed is made up in a granular form, then the birds will still be able to scratch and find it. It doesn't matter if it has a bit of litter stuck to it, as it will still be eaten.

Meal and mash

Firstly, there is the meal as it comes straight from the mixer at the mill. Because of its fine texture, it takes the birds longer to pick up enough to satisfy their appetite and therefore longer to eat, so they will have less time to feel bored and will be less liable to vices such as feather-pecking. One disadvantage, however (and this is if it is a coarse milled mash), is that the birds try to pick out their choice morsels, and in doing so, flick the other components of the mash out on to the floor, mixing it with the litter and wasting it. This problem can be alleviated to some extent by having a spill-proof grill and a lip on the feeder.

Mash (a mixture of ground grain and nutrients) can be fed to birds of all ages but will come in different compounds with a varying protein content. Your chicks—if you are rearing from day old—will need a high-protein content if they are to grow quickly and efficiently. When they are about five weeks old they can be fed a rearer's ration, which contains less protein and is less expensive. Just before they come into lay they can be moved on to a layer's ration, which will induce the production of eggs, and you can continue to use it throughout the laying cycle. If your hens have been fed the correct foods throughout the growing time their bodies should be able to withstand the stresses of laying hundreds of eggs per year. All of these changes from one form of food to another should be done gradually. Mix one food in with the next for at least three or four days so that the birds will continue eating and not go off their food.

Pellets and crumbs

Young chicks eat very small pieces of feed, or crumbs as they are called, that they can pick up in their beaks and swallow. To encourage them to seek out food, hard boil a couple of eggs, mash them, then add them to the crumbs. The bright color of the egg yolk will catch their attention and they will start to eat, like the taste, and you should be well and truly set on your rearing program!

Crumbs are produced from pellets, a form of feed used to feed both growers and layers. The mash is taken through a machine in which steam is added and then forced through a metal surface with small round holes to produce strings of mash that are then broken down into pellets small enough for hens to pick up and eat. For chicks, these pellets go through a further process to break them up into even smaller fragments, known as crumbs.

Grains

When grain prices were much lower, chickens were fed unprocessed grain. Most grains are very palatable to hens and have the advantage that, when scattered in the litter, the birds will scratch for them and therefore turn the litter over for you. With grain prices doubling and no decrease expected in the foreseeable future, the only real advantage of cheapness has all but disappeared. Still, whether it is tradition or not, many still choose to feed grains to their hens.

Modern mixed corn is usually mainly wheat with added maize, oats, and barley. Some suppliers now mix in a little soya oil, too, so that there is more vegetable protein. If you live in the countryside and are near a farmer who grows wheat you may be able to negotiate a deal to buy it direct at the same or similar price paid by the grain merchant.

Maize is grown mainly in North America, and although highly palatable and useful for keeping up the color of egg yolks, should be fed sparingly; too much and the bird will pile on fat, as maize is a very good source of energy and will help keep the bird warm in winter but can halt the laying process by coating the hen's reproductive organs in fat.

Oats are less palatable than wheat and maize but have many beneficial effects, including a reputation for building a strong frame and helping with digestion due to its high-fiber content.

Barley is the least palatable to poultry and is therefore used sparingly and mainly in rations for other animals, particularly pigs. Most chicken keepers agree that if you give a feed of grain just before the bird goes to roost it will quickly eat it up and will then spend the night with a full crop of food working through the digestive system. This means that the bird will not feel hungry through the night and its digestive process will be making the natural nutrients for producing the next day's eggs.

If you do have a source of cheap grain available to you, spread some in a shallow tray and pour over a little water. If kept at a temperature of around 59°F it will soon sprout and grow. Then, when it is about 2 in. tall, feed it to the birds as a treat and they will relish it.

Ingredients and quantities

The ingredients of bird feed are always listed in descending order of weight (so the heaviest ingredient is listed first). For instance, the analysis of a typical bag of layer's feed would read as follows: wheat, followed by wheat feed, soya, calcium carbonate, peas, maize, grass meal, beans, and linseed. Most will also contain dicalcium phosphate and sodium chloride.

A recommended daily ration per bird during its laying cycle is 3½–5⅓ oz. of feed per day. If you do, however, choose to feed grain, then it is recommended you feed additional calcium in the form of limestone or oyster-shell grit to help with the grinding process. This can be bought in the form of mixed grit where flint grit is also mixed in.

The grinding process

This is part of the process by which poultry birds digest their food. It takes place in the gizzard, a circular organ with flat sides that forms part of the digestive system. The gizzard contains ribbed muscles that, with the rough food and the mixture of sharp grit, grinds the food down into a readily digestible mass to pass into the intestines. This is particularly necessary when the bird has access to grass. There is even a very fine grit that can be fed to chicks. If you cannot find it under the heading of "chick grit," you should be able to locate it among the feed for cage birds.

Household food scraps

Can you feed household food scraps to your birds? In short, your poultry birds can be fed some, but not all, household scraps.

During the war years when poultry feed was hard to come by, chickens were fed potato peelings and other vegetable peels, leaves from plants in the garden, and the like, all boiled up and drained off with a form of meal. Not so many years ago most poultry feeds used animal protein in the form of fish meal and it made the feed very palatable; many of the vegetable proteins were not as tasty to the hens.

All vegetable peelings must be boiled first and you now cannot feed your birds meat or fish scraps. This is because of the risk of transferring diseases that might be found in undercooked or raw meat or fish that could then find a way into the human food chain via your chickens. Government legislation forbids it, but whether a government body would forbid you feeding your table scraps to your own birds if you alone are consuming the produce is another matter!

However, most vegetables contain protein, which is essential to the production of eggs, growth of new feathers, and general health. Potatoes, however, have a high proportion of water, despite also containing starch. Bread, which can be dried off and used to mix into cooked scraps, usually contains around 10 percent protein. If you are hoping to save money by feeding your birds scraps and dried bread, you may be out of luck, however. By the time you take into account the energy used to cook your waste, the time you spend preparing it, and the fact that, especially with hybrid hens, you will lower the number of eggs you can expect to receive from birds on such a diet, you are actually better off buying ready-made chicken feed. You may, however, gain some satisfaction from the fact that you are no longer wasting these edible materials. In the evenings, instead of feeding your birds expensive mixed corn, you could, by using layer's meal to dry off your cooked mixture, produce a feed of similar nutritional value but at a lower cost.

If you grow your own green vegetables, use the hard stalks from cabbages, cauliflowers, or sprouts that you would probably discard to feed to your birds. Hang the raw stalks up by the roots at a height that the birds can comfortably reach in the run or hen house. This will provide a tasty variation from which the birds will enjoy picking. Do take time to hang up this plant material, as if it is left on the ground and soiled the hens will soon lose interest. When almost finished, split the stalk and they will eat the center, too.

Troughs and hoppers

How do you dispense these various types of feed to your birds? The type of feeder you use mainly depends on the number of times a day you intend feeding your birds.

A long, low trough with sufficient space for all the birds to feed at the same time is the best option if you are feeding twice a day (mornings and evenings). It is not necessary to feed more often than this.

You can take this opportunity to monitor whether all of your birds are eating. Maybe one will be in the nest box laying, but, if it is reluctant to come out, it is worth keeping an eye on this bird in case it is unwell. Feeding twice a day has the added advantage of deterring rodents, who will always be attracted if food is left lying around, even for reasonably short periods.

A hopper-type feeder, often of a cylindrical shape, is usually used if you do not have time for a morning feed. The shape of the hopper means the contents will fall into the trough at the base, ready for your birds to pick at. The food trough or hopper should be placed inside the house to keep it dry. As long as you make sure you regularly check that the contents are not running low, then all should be well with your birds.

If birds run out of food for too long and too often, it will play havoc with both their growing and laying. Without regular and reliable access to their feed, both processes—growth and laying eggs—will come to a halt. You could also find your birds indulging in vices such as feather pecking (see pages 65–6) and, in the extreme of being kept in an area with no access to natural food, cannibalism.

As with most problems, the old adage is first to look to the source. When buying one of these feeders you should make certain it is right for the chickens. Some poultry keepers still like to have their feeders outside the house but within the confines of the run, although government bodies recommend otherwise. The thinking here is that wild birds can transmit disease and you should not encourage them to mix with your flock. Outside feeders should have a removable cover that, when in place, will shelter the feed from rain or bird droppings. Wet feed, if left static for any time, soon becomes moldy.

The hoppers you use in the house are best hung up so that rodents cannot steal the food and contaminate it. It also prevents the birds scratching their litter into it. If it is hung on a cord it is usually too unstable for the birds to perch on.

Water

It is advisable to place the water container outside in the run so that the house will not get wet through any spillages. As the birds are closed in the house at night to protect them from predators, they will not have access to water. But do not worry; birds spend most of the hours of darkness asleep and do not need water. If you are still concerned, however, then a slot about 6 x 2 in. cut through the house into the run with access to a water container should do the trick. Place an outside drinker in the run next to a door that you can easily reach. Some drinkers, especially those of the modern fount design, nearly always have a handle at the top, so if you are having trouble with the birds turning the drinker over, secure it in place with the aid of a hook positioned through the roof of the run.

Storing food

Wherever you store your food make sure it stays dry and is protected from rodent attack. Kept in a brick or stone building it should be safe, but it is recommended that as an additional safeguard you use a metal container. This can come in the form of a purpose-built feed bin or a metal dustbin. A rat can very easily chew through a plastic bin.

Make sure your bin is completely emptied and cleaned out on a regular basis, prior to filling it with your next purchase of food, otherwise little mites will make their home in the residue and contaminate all around it. If you buy your feed in paper sacks, as most of us do nowadays, then put however many sacks it will accommodate into the bin and when you have used all the feed from the sack it can be thrown away, burned, or recycled. You will know then that all the feed you use is fresh.

Use a vacuum cleaner to clean the inside of your feed bin; with its many attachments you should be able to reach any awkward corners or edges. If you decide to wash it out use an government-approved disinfectant but follow the recommended mixture and leave it to dry out thoroughly and allow any odors to reduce to a minimum, or you may find it flavors the feed and consequently the eggs that will be laid by your birds.

A final word on your food storage. As the bags of feed you bring home from the supplier are likely to be heavy, store your stock as closely as you can to your transport. If you cannot carry the bags very far a wheelbarrow will be useful. As with all food nowadays, bags of feed have a use-by date. If it is a few weeks over ask the retailer for a discount, as the nutritional value will only be reduced by a very minimal amount.

Bringing your birds home

At last we have gotten to the point of the whole exercise and you should be ready for your birds' arrival! Make sure that when your hens arrive you are at home all day to see that they settle properly into their new surroundings.

We will assume you have chosen the most popular option of point-of-lay pullets. If you are collecting them yourself, here are a few tips that will make transporting them successful.

Transporting the birds

Firstly, take note of the weather. If it is cold you should not have too many problems, especially if you are not traveling too far and have suitable containers for carrying your birds. Pet carriers with plenty of air holes are ideal. You could also use pet-carrying boxes but it is best to cut larger holes near the top of them for extra ventilation. If you are using cardboard boxes make sure the bases are strong enough and that you do not try to squeeze too many birds into one box, as they generate a lot of heat when crammed together.

Many high-street retailers throw out strong cardboard boxes, so look out for these; you only need to ask and they will give them to you. Make sure you cut enough air holes near the top to let out the heat. Put some form of litter in the bottom: shredded paper, straw, or hay are best. Without litter the birds will be thrown around the box as you drive and will get very stressed and may even injure themselves. If you use shavings or sawdust you may find a fair proportion of it scattered over the inside of your vehicle at journey's end, so cover the car seats with a protective sheet.

In cold weather you can carry your birds in the trunk of the car, as long as it has some air circulation. If the weather is hot and sunny, make sure the sun does not shine directly onto the box, and under NO circumstances should the birds be carried in the trunk of the car in hot weather.

Unless you are traveling a very long distance to collect your birds, there should be no need for feeding and watering during the journey. The motion of the vehicle will mean that the birds spend most of their time sitting down and will not be interested in food or water. If you are traveling a long way and feel that you need to feed and water them, take the carrier out of the car first so that spills do not occur inside. If you are carrying your birds inside the car in containers with lots of ventilation holes, an old bed sheet or something similar placed under the box could save a cleaning job later; there is no guarantee your birds will not empty their bowels through one of the holes in the box.

Upon arrival

When you reach home place your birds in their house. Make sure they have food and water, and leave them to settle down without disturbance in their new surroundings. Unless it is very early in the morning it is a good idea to leave them inside the house until the next day. There are a number of reasons for this. Your birds will probably not have been handled very much, nor put into carrying boxes or transported, so a nice quiet house where they can settle down and investigate the food, water, and perches will be welcome. If, once it is dark, you find them on the floor of the house, try placing them on the perches, where your birds should go to sleep. If you are not successful after a couple of efforts, then leave them until the next day.

When you handle your birds you may be able to feel if the crop on the front of the breast has food in it. If the birds have been eating pellets or corn you should be able to feel the granules through the feathers and flesh.

Next morning, carefully lift the chicken hatchway so as not to startle the birds. Opening the chicken hatchway will give them the chance to go outside and explore their new run. As the surroundings will be new to them they may not come out right away. Remember, they are in a strange house so they may not be aware that the door opens. If all appears well, leave them to make their own way out. If you can spare the time, return in about an hour or so and you will probably find them outside, scratching around.

Immediate care

If you are feeding the birds twice a day, get them used to a routine of when you will appear with the food. Should it be, say, half an hour before dusk, then feed them in the hen house, scattering a little by the chicken hatchway to encourage them to go in. Hopefully you will not have to catch them to put them in but, if you do, it is best to leave it until after dark, when the birds will be sitting quietly and you should be able to collect them without too much fuss. A hand placed on either side of the bird, holding the wings to its body and the head towards you is the most successful. Stand them with their feet on the perch and once they have a hold let them go gradually and hopefully they will stay there. It is wise not to try to chase them into the house, since this will frighten them; they will only respond to being "shooed" in once they already know you well.

Buying your birds at point of lay means that when you get them they should be fully grown and have all their adult feathers, but the comb on the top of the head will probably still be small and not yet bright red. As they get closer to lay, they will start to talk to you much more and the comb will become bright red. If a bird is not out in the run, you should not be in too great a hurry to check the nest box, as when they first start laying they will be nervous. Quite often once they have laid they will announce to the world what they have done as they leave the nest, and you should then be able to collect your fresh egg. Startling them if they have laid but are still in the nest may result in a broken egg. If this happens clean it up right away so they are not tempted to eat it. Once birds get into this bad habit it is very hard to break them of it.

Baby chicks

Do take extra care when buying baby chicks and try not to get too carried away! Newly born fluffy chicks are gorgeous to look at and feel soft and warm, and many people are tempted to buy them. But chicks require a lot more care than point-of-lay pullets and if you are unprepared for this level of commitment, they may not survive.

Before your chicks can be brought home you need a few extra essentials. A large wooden or cardboard box or a cage lined with shredded paper or wood shavings to keep them in for a start, preferably indoors away from cold draughts. You then need a heat source, and there are plenty to choose from. You can get away with a 60-watt light bulb hung above one end of an open box so that the area under it is warm enough to keep the chicks happy. Leave the light on at night, as you don't want your chicks to become chilled. The more professional option is to buy a dull emitter heat lamp that will provide enough light for the chicks to find food and water.

How do you know if the temperature is right? The chicks will tell you! If they huddle under the lamp and make cheeping sounds it means they are cold. In this instance, lower the lamp a little so that it is closer to the top of the box and provides more heat. If the chicks lie down in a ring under the lamp and do not call out, you have got it just right.

The part of the box underneath the heat source should be covered with a sheet of plywood or plastic. Do not use tin to cover the box, as it is not a good conductor of heat. The heat will not become hot enough to create a fire hazard, so do not worry. Make sure you do not cover the entire box, as the heat generated will build up and can be fatal. Food and water should be positioned at the other end of the box, away from the heat source. Once the chicks have fed and watered themselves, they will naturally return toward the heat source.

We have already learned about adding mashed hard-boiled egg to the food to attract the chicks. To teach them to drink water, hold them gently and dip their beaks in the water and they will soon realize what it is. You only need to do this for one or two of the chicks; the others should then follow their lead. If you are feeding the chicks hard-boiled eggs, never make so much so that it sits in the feeding dish for too long, as it will very soon turn sour. If they trample it flat just mix it up with the crumbs or mash and it will be easier for them to eat.

Most crumbs and mash feed bags will feature the letters ACS, which is short for anti-coccidiostat. This is an additive that protects the chicks against Coccidiosis (see page 64), an infection that can be fatal. It is usually caused by damp, dirty conditions, so always change their litter regularly when they are young—as soon as it gets wet or soiled. Clean up any water spillages as soon as you can and put down fresh litter.

For the first few days place the chicks on paper towels; this gives them a surface with a better grip than a smoother lining, such as newspaper. The paper towels should be changed every day. Chicks can get splayed legs if they are sliding around before their muscles are fully developed. One indication of Coccidiosis is red droppings. If you see red droppings give them fresh litter as often as you can, as they will be attracted by the color of the droppings and will peck at them. This may sound horrible but it does happen and can prolong the infection cycle, which comes round every forty-eight hours.

As your chicks start to grow and get some feathers they will need less heat. Some heat lamps can be adjusted in terms of the amount of heat they give off. If you are using an ordinary light bulb, then you can lower the heat by raising the lamp higher above the top of the box. Never do this last thing at night, however, as you will not be able to assess if the chicks are still warm enough. Unless you keep them inside a solid building the temperature will naturally fall during the course of the night. Gradually lowering the heat does promote feather growth and you should not need any heat after six to eight weeks, depending on the weather. As the birds grow, remember to increase the size of the area available to them.

Overcrowding can lead to some chicks receiving all the heat and much of the food, while others will become weaklings. Until your chicks are off heat, unless it is summer time, it may be wise not to allow them access to the outdoors. When you do, make sure they are restricted to an area near the house for a few days. By the time they are six to eight weeks old they will have grown most of their young-bird feathers, depending on the breed.

Growing stock

The fact that birds between brooding and laying age are not only growing but also getting their feathers means that it is essential that you maintain the quality of food you give them. Feathers are made up of plenty of protein and as they grow blood circulates through their stems carrying all the nutrients. Growing is a steady process, not only for the body but also for the feathers. The initial covering of feathers that your growing bird acquires will change before it comes into its adult feather, prior to laying.

The food you feed your birds during their growing, and just prior to their laying, periods needs to be different. It is a fact not widely known that the development of the yolk begins around eighteen days before an egg is laid. To produce an egg a hen needs a diet that is high in protein to produce the contents and minerals needed for the shell. During its growing time, however, it needs less of these and more of the carbohydrates for energy, so to get the full benefits of the laying cycle make sure you feed the correct feed. You may be able to tell when the yolk begins to develop by the look of your bird. Eighteen days prior to laying, the head, comb, and face become brighter red in color and the fluffy area at the rear of the bird becomes fuller. You then need to put your bird on to a wholly layer's ration rather than a grower's ration. This will give the bird all the nutrients it needs for producing its eggs.

As young chicks your birds will have spent their time under the heat lamp and living on the floor. As they start to acquire their wing feathers they will try to fly. If they can they will find somewhere off the floor to perch, not only at night but also during the day. Do not encourage them to perch until they are about eight weeks old, as their bones are still soft and they may end up with deformed breast bones, or, with heavier birds, damage to their legs. It is wise to set low perches at an early age, lifting them up as the birds get older. Make sure your perches have a flat, fairly wide surface, not only to protect against deformed breast bones but also breast blisters. Of course, it is also more comfortable to sleep on a flat surface.

When moving your young birds to a new house if they have not been accustomed to perching it is wise to check at night that they are not suffocating one another. This can sometimes happen if they all gather into a corner and climb on top of one another. A short piece of board positioned across the corners can help but, as with adult birds, it is a good idea to go out after dusk and place them up on the perches. This procedure should only be necessary for the first few days.

Unless you have water always available, your birds will not initially grow or later lay eggs. Without a ready supply of clean, fresh water they will not be able to process their food. When it comes time for egg production water is doubly necessary. It has been found that lack of water not only leads to vices but also smaller eggs, even if the birds still continue to lay. Most drinking containers will not need to be filled more than once a day.

Essential routines

After the initial excitement of buying your first birds and settling them into their environment, you must accustom them to a routine of care. Most of this will be common sense but it is worthwhile setting out the chores that will need to be carried out on a regular basis.

Daily tasks

Although you need to be very committed to keeping chickens, it need not take up endless hours of your day. Once your birds are settled into their environment it is up to you how long you spend with them. Basic chores consist of providing enough food and clean water, and closing the birds in at night. The birds will automatically settle in the house once it gets dark, and this is a good time to check they are well provided for the day ahead and to collect any eggs. Chickens lay during daylight hours and you should always try to pick up the eggs daily in case any get broken or if one of the hens decides to sleep in the nest and fouls the eggs.

In theory you could spend just fifteen minutes at the end of each day attending to the basic requirements of your chickens, though a thorough clean-out is necessary at least once a week and this will take longer. Unless you were pushed for time I would also expect you to spend a little time observing your birds. It is only when you take time to study your hens that you become aware of their behavioral patterns. Humans have good and bad days, and so do hens! If a bird is reluctant to leave the house it may be laying or it may be out of sorts. You will generally notice any unusual behavior on your first visit of the day. If the problem lasts for more than a day you may have to check for a number of ailments that affect hens (see pages 59–64). Do not immediately rush off to the vet; it can be expensive and many are not experts on hen ailments.

If you have a moveable house and run the number of times you move it will depend on the weather and the size of the run. Fresh grass, so long as it is not old and tough, is a very good daily tonic for your birds. If the house and run are separate it is easier if you move them when the birds are in the house. It may be heavier but can be less difficult than chasing birds around the garden! If they do get out, a handful of food thrown in the run should entice them back in.

Feeding and watering

Feeding will occur naturally when you visit your birds. If your feed is in a sack, then a bucket and a scoop are all you need. If you do not need much feed, then just use the scoop to put the feed in the trough or feeder. If your birds collect around you, be careful where you place the bucket or they may turn the contents over and make a mess.

If some droppings have got in the feed trough, do take the time to tip them out. The same applies to the drinkers, and probably more so. Dust and litter will collect in the part of the trough where the birds drink. Not only will this contaminate that area but also the inside of the drinker. You should regularly empty the contents and scrub the inside and the drinking area with a brush, checking at the same time that the holes where the water comes out of the container have not been dirtied by any residue. Also check that the water container does not have any leaks. It is only too easy to think you have cleaned everything up to find next time you visit your birds they are desperate for water.

Collecting eggs

Feeding time is also probably when you will be collecting any eggs, so check the nest boxes regularly. Make sure you do not put the eggs anywhere where they can be knocked off and broken. If you have dirty eggs and have to wash them they will not stay fresh as long as unwashed ones, as washing removes the waxy cuticle that covers the shell, leaving the shell's pores open and the contents susceptible to evaporation. You can never guarantee your birds will not carry some dirt into the nest box and some may even sleep there at night, therefore a regular check and changing of the litter in the nest box is essential. Some modern bedding materials do have an odor and should be avoided as it can make the eggs taste unpleasant; eucalyptus-flavored eggs are not very appetizing!

Cleaning

Remove any wet litter and accumulated droppings from the house at least once a week, but if you have time on your hands you could clean out the area once a day. However, such frequent cleaning means that you would amass much more waste litter that will take longer to break down as it will contain fewer wet droppings. Do not throw this litter away; it can be added to your compost heap or sprinkled around the garden.

The litter you use on the floor of the house will govern how you make use of it in the garden. Litter such as chopped straw will, when cleared out and mixed with a fair amount of manure, break down fairly easily into a mulch you can use in the garden. Shavings and sawdust will need to generate a fair amount of heat before they break down into a usable mulch. They can, if used too soon, make the soil very acidic. This is not bad if you have shrubberies where you grow acid-loving plants, such as rhododendrons and azaleas, although you should not make the mulch so deep it becomes moldy.

If cleaning out litter from a large run that is not moveable, make sure you collect any larger accumulations of droppings. These will likely be found around the house and gate where you enter with the food, as the birds tend to congregate around these areas. The remaining areas will only have odd droppings here and there. Although a small amount of litter can soon be incorporated into the grass, too great an amount will kill off the grass. If this happens, you can soon have the area green again by scattering grass seed and lightly raking it in—but make sure you seal off the area you are sowing or the birds will scratch up the seeds.

When you have finished cleaning up the house you should check the run. If you have a large permanent run, then the level of maintenance should not be too great at the start. It is worth, however, regularly checking that all the fixings that hold the netting to the posts are in place, and that the enclosure is predator proof. If the netting starts to lift up from the bottom—which could allow a fox to dig under—either cut some wooden pegs or lay a heavy piece of metal, concrete blocks, or bricks on the problem areas to anchor down the netting. Vigilance is a necessity if you are to avoid any problems. Once the grass has grown around the base of the wire it is not as easy for a predator to push under. If you live in an area where foxes roam you will soon learn to recognize the sound of the male barking or the vixen screaming, or even more obviously when you become acquainted with it, the scent they leave. Badgers can also be a problem but that is usually quite rare.

Other routine inspections

If you have cut a heavy crop of grass, do not leave it around to heat up and rot; clear it away or add it onto a compost heap, if you have one. Similarly, if your chicken runs are short of green stuff you can feed them small amounts of short lawn clippings so long as it is only enough to be consumed that day.

Sometimes, and this usually only happens in older birds when they are not scratching for worms and pecking grass, you may find the top mandible of the beak grows longer than the bottom. Not only does this look parrot-like but it can also be inconvenient for the bird when it tries to eat. Whichever way you choose to cut back the top mandible do not be too severe. The best instrument to use is nail clippers. Do not cut straight across the end as this may split the beak up through the

center. Instead cut diagonally from either side, preserving the point. You may also have to do this on toenails but again don't be too severe or you can make them bleed.

Observing your hens

After all these troubleshooting routines the best part of keeping hens is bird watching. Unless you have had the chance to watch someone else's hens or talk about them, this is when you learn to understand what makes a hen tick. A hen is a creature of habit and although it has a brain, it is very small, literally the size of a pin head. Many of the old sayings from the countryside are based on the habits of our creatures: "The early bird catches the worm" is vividly illustrated in the rush of any type of poultry when released from their house.

As you get to know the birds as individuals they will become more interesting. Hens have their own personality and do not usually all act in the same manner; some will be confident and others more nervous. The easiest way to attract them to come close to you is to place a bucket of food by your side and watch who leads the rush for it. Quite often the most forward is the "bright-looking" bird that produces the most eggs. A bright-looking hen will have a bright red comb and bright red/orange eyes.

At the start of the laying cycle your hens should all have soft, smooth plumage. However, as they continue laying the plumage starts to look a little frayed. If you find one of your hens always looking in top condition it could be that she has been looking after herself rather than producing eggs for you. Not all hens lay regularly, so a handling test of their laying capabilities will give a definite knowledge of what they are doing at the time. Catch your hen by placing one hand on its back and the other on its side and then removing the hand on the back and placing it on the other side of the bird. If your bird is tame you may be able to catch it by placing a hand on either side. Pictures of commercial people carrying chickens by their legs can be seen but this is not the recommended method. Once you have caught the bird, turn its head towards you and lay it along your arm with one hand holding the top of its thighs. Place the other hand on the area below the vent where the fluffy feathers are.

If you want to work out if your bird is laying, place four fingers horizontally between the pelvic bones and the rear point of the breast bone. Then try to place your fingers between the pelvic bones; if the bones are hard and there is only about one finger's width between them, then the bird is not laying. However, if the bones are thin and pliable with around two to three fingers' width between each bone, your hen is definitely in lay.

Egg facts

The fact that you can now have your own hens and can pick up an egg as soon as it is laid gives you a satisfaction you could never experience when buying them "fresh" from the supermarket. How can you tell if an egg is fresh? The following tips will guide you:

- It is always harder to peel the shell off a hard-boiled, newly laid egg than one that is four or five days old.
- When you break an egg into a frying-pan, if the albumen (white) spreads all over the pan, it is old. It should hold together and the yolk should stand up.
- There is a theory that the contents of darker-shelled eggs do not evaporate as quickly as those of pale-shelled eggs, the idea being that the extra layer of pigment on the darker shell protects the pores.

The magical molt

The arrival of autumn brings with it a noticeable physical change in your chickens. The birds will molt once a year and this usually occurs in the autumn. The annual molt is when the hens stop laying and change their feathers for new ones. The length of the molt depends on the bird, as some drop all their feathers over a short period of time and others will do it piecemeal and take five to six weeks. The birds will start laying eggs again after the molt.

You will be expecting each point-of-lay pullet you bought to lay about 320 eggs in a year. If your birds are hybrids, and you bought them in the autumn, gave them extra light during the winter, and had no other problems, then these birds will probably not molt until the following autumn, twelve months after you purchased them. If, however, you bought them in February, for instance, and in the autumn they stop laying, then the molt has started. You may notice your pens and houses are covered in feathers and the birds take on a very untidy look.

The significance of the molt

Eggs and feathers, when they are growing, are both rich in protein, and your hen cannot produce both efficiently at the same time. Eggs contain around 13 percent protein but new feathers contain much more, about 80 percent. You will probably be aware by this stage that most layer's feed contains around 17 percent protein, so the bird makes the choice whether to put this protein from their diet towards egg production (during the laying season, which is most of the year) or the molt (which usually occurs in autumn).

Good laying hens' feathers become frayed during the laying season. With winter just around the corner the hens feel the need to get a new, more efficient covering for their body against the imminent wet and cold weather. This could almost be compared to applying a new coat of paint on a house to protect it from the weather. So the annual molt is simply Mother Nature looking after your hens.

A natural reaction of many newcomers to poultry keeping is that when their birds stop laying they give up feeding them pellets or mash and opt for a cheaper alternative instead, such as straight wheat or barley, which will be lower in protein. This could be a false economy, however, because keeping the hens on the higher protein food will bring them back into lay more quickly. Molting is a time of stress for your hens and so caring for them at this time should be a priority.

Starting to lay again

As molting usually occurs in the autumn, when the molt is just about complete, it is necessary to give your hens a helping hand back into egg production, otherwise they may not start laying again until early the following year when the days start to lighten. An easy way of accelerating this process is to artificially lengthen their hours of daylight with the use of a standard light bulb on a timer inside the hen house. The timer is best set to come on early in the morning, otherwise you will need some sort of dimmer if you want to use it in the evening. The reason for setting the light to come on in the morning is that if your hens are on the floor of the house when the lights go out they will have no way of finding their way up on to the perch for the night. The optimum amount of light hours is around fourteen hours per day. Light stimulates the pituitary gland, which in turn influences laying.

Winterizing your birds

Hens love to run around outside but during the cold winter months they will need a little extra care from you. Unlike humans, hens can't put on an extra coat for protection so you must take measures to protect them from the cold. Cold, unless it is really intense and for long periods, need not be detrimental to your hens. Winter winds can be held back by fitting wind-breaks around the run where they spend most of their time. Backing the hen house on to the north and having the daytime sun warming the other side will benefit them. Keeping the inside of the hen house dry is always very important, but even more so during the winter months. If there is snow with a driving wind, use some form of cover to divert the snow away from the chicken hatchway or windows. Do not block them up completely though, as this will make the house stuffy.

You may, at this time, add a few extras to the feed. One of the best additives used in countries where long periods of freezing weather occur is maize. This is a very good source of energy and will also put an extra layer of fat on your birds. Don't feed your hens too much maize (no more than one eighth of their feed ration), however, or it will slow down the laying process. Cod liver oil or various vegetable oils mixed in with the feed will also help the feathers retain their protective qualities during the cold winter months. There are also a number of other additives that are claimed by their manufacturers to have benefits. If you wish to try them out yourself, do make sure they warrant the extra expense.

Another factor to consider is the water supply. Not many hobbyist poultry keepers will have, or need, automatic water systems, which in cold weather need constant monitoring. These systems are usually only really used in intensive indoor rearing. If the weather is particularly cold it may be best to empty drinkers in the evening and fill them the next day, to prevent the water freezing overnight. It is hard to break or thaw out ice from a drinker and if you have a plastic drinker it is liable to split on impact, being more brittle in the cold. There are some proprietary heaters that can be used to keep the water temperature above freezing and these are similar to those used in outside fish ponds. During very cold periods you may find you have to fill drinkers twice a day. If you get home late and don't have electricity in the hen house, a dim lantern will enable the birds to see sufficiently to go to get a drink and then get back on to their perches (you don't want to confuse them into thinking it is daylight by providing too bright a light at night).

The benefits of second-year birds

In most of the commercial world hens are only kept in production for one laying season. There are, however, benefits if you are prepared to let your hens rest from laying while they are molting. They may be costing you money with no return but you must also consider the benefits when they start laying again. Because you are keeping them a second year, you do not have the expense of buying new birds. A hen will lay on a decreasing scale for around ten years, if it lives that long.

Your hens may not lay quite as many eggs in the second season as the first, but on average the eggs they lay will be larger. They will also have settled into their environment and know one another, so hen pecking should no longer be an issue. This should be true for all hen types, whether hybrid or pure bred. When it comes to future years it is generally found that a hybrid's performance tails off, as it was originally bred with the aim of laying the largest amount of eggs possible in the shortest time on the smallest amount of food. With most pure breeds, however, laying has always been a required characteristic, and many breeders have kept their hens on for longer periods because they are not solely governed by the profit margins on their hens' produce.

The end of laying life

What you do when your hens stop laying will depend on your outlook on profit or sentimentality. Most small poultry keepers will let their birds live their life till its natural end; others may wish to replace their old birds with new ones. If you wish to keep old hens on, then you may either have to introduce new hens into your flock—if the house and run are large enough—or you will have to consider investing in an extra house and run.

In the commercial world of old, when hens reached the end of their useful laying life, they were killed off and sold as broilers. It was no use trying to roast them as you would a young bird because leaving the bird in the oven for too long in an attempt to make it tender only dried it out; putting it in a large saucepan with some water, and steaming or boiling it, make it edible. The fact that it was an older bird, however, meant that it did have more flavor. In the commercial world nowadays, old hens get processed into animal food, glue, or other products.

Killing and plucking

The most popular clean method used by small poultry keepers of killing poultry is to dislocate the neck. Until you have had this demonstrated to you by someone who knows what they are doing, this is best left to an expert. It is quick, efficient, and painless if done correctly. After killing, plucking is an art that, if done properly, does not tear any skin and can leave a professional finish. The feathers come away from the body more easily if it is done when the body is still warm. The best way is to pull the feathers in the direction of the head, against the way they lie on the body, holding them nearest to the base.

The hardest feathers to pull are those on the wings and tail and it is best to pull these sideways and singly. Use a pair of pliers if you are finding the job too difficult with your fingers.

Preparing for the oven

When aiming to kill a bird for consumption, it is always best to keep it off food for at least twelve hours before slaughter. This means that the main organs will not be full of food when you attempt to remove them. After plucking, start by removing the legs where the scale section joins the flesh. Then cut the skin around the base of the neck, and cut and twist the main section off. Clean out the crop from the front of the neck and insert your forefinger through the cavity into the body, and with sideways motions loosen the lungs from the backbone. Then make a small cut at the back of the bird between the breastbone and the vent. Open the bird up and draw all its internal organs out. Finally, cut the skin around the vent and carefully clean it away. A quick rinse under cold running water and your chicken is ready to cook.

I hope this is not seen as a gruesome exercise but rather a reality of everyday life. These methods can also be used if you hatch your own chicks and rear cockerels or find that you are overrun with chickens. Again, reality and common sense dictate that the cockerels must go. Pure-breed poultry keepers will often let their male chicks live a little longer to see if they make the size, shape, and color that is desired of the breed for exhibition purposes. Unless someone comes along who wishes not only to have eggs but also breed for the future, the breeder will have to dispose of any extra male birds. The value of a dead rooster bird for the table is never as much as a special one sold for breeding.

Breeding your own

Although the primary motive for keeping chickens is the production of your own eggs under humane and attractive conditions, it could be that the idea of having a rooster in your hen pen in order to breed your own replacement birds, build up your flock, or supply other prospective poultry keepers with their first hens may appeal to you. The amount of time and money spent on this process will, in most cases, be more than the amount you would spend simply buying replacement birds if all the costs such as labor, heating, housing, and feed are taken into consideration. However, much satisfaction can be gained in producing your own laying hens.

This will only be possible with pure-bred rather than hybrid stock. Parent stock for hybrids are bred specially and only franchised out to licensed breeders.

It is a fact that the male bird, with his bright shiny colors, is usually more attractive than his female counterpart. A group of hens being called by a rooster to some choice morsel he has found is quite a sight—they usually run to him when he calls. When a male bird is present the hens tend to collect in a group around him. He will often be on the watch for any problems, such as predators, and will give out a warning call. As with hens, there are different

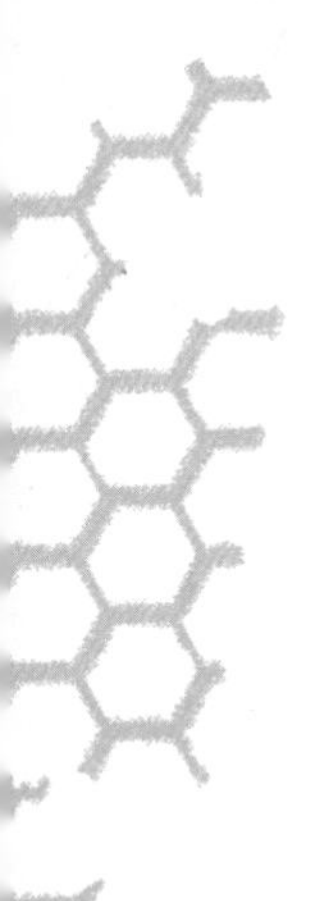

names for a young and adult male. Up until its first molt, usually around twelve months, the young male is known as a cockerel, after that a rooster. If you are breeding from them it is better to use a young male more than six months old to ensure better fertility results. Cockerels are always more vigorous in their mating habits than roosters and this will sometimes show in the feathers being pulled from the females' backs. Such damage should not occur unless the rooster bird is very old and the spurs on the back of his legs are very long and sharp. If the spurs do grow too long (this is more common in some breeds than others), using a sharp pair of strong clippers to take off the points should save severe damage. It does help to have sufficient hens to keep the male entertained—five or six should be a good number. This gives the rooster plenty to choose from, although he will usually have a favorite he chooses to mate with more than the others. Too few hens may result in them becoming stressed and this can sometimes curtail their laying.

For mating purposes, put male and female birds of the same breed and variety together, as cross-bred birds are not as much in demand. You will then know what your chickens will grow up to look like. Introduce the rooster at least two weeks before you start to collect your eggs for hatching. It is likely that, for the first few days after introducing the male, egg production will diminish, as his presence will have interrupted the hens' normal routine and way of life. If you are using an old rooster for mating purposes, then he may not be fertile too early in the year. The saying is "Maybe he needs some sun on his back!"

If you are worried about your rooster crowing early in the morning, then there are a few steps you can take to reduce this. Soundproofing the inside of the roof of the house with polystyrene materials is a good option, as sound usually resonates through the roof more than through the walls. You can also pick the rooster off his perch at night and put him in a ventilated box in the garage or a separate outhouse. Do

make sure, though, that it is dark. Switching on the light will often result in him waking and crowing. It could be that the only person the crowing affects is you, and in these days of double glazing his crow may not wake you up. If in doubt, do ask your closest neighbors if they are being disturbed by the crowing. It is worth noting that some of the smaller bantam breeds have a much more shrill call than the large heavy breeds.

Dealing with a troublesome rooster

Sometimes, for no apparent reason, a rooster may become aggressive towards you. Initially, this usually takes the form of him standing in front of you with his neck outstretched and hackle feathers raised. If he does not attack you, ignore him—kicking out will usually only encourage him to fly at your leg. If he persists with this behavior, and if he has been tamed beforehand, then pick him up. He may then see you as the friend you are. Sometimes wearing bright-colored trousers or footwear encourages attack. One thing you should not do if he persists is to turn your back on him, as he may fly at your legs and cut them with his claws and spurs.

Should he persist with this bad behavior (carrying a stick to defend yourself does not work) the only solution is to get rid of him. However, you should only give him to someone else if you warn the new owner of his vice. Sometimes it happens that he will only attack female poultry keepers and leave men alone. The time when a rooster is most likely to misbehave is in the springtime, when his hens are starting to lay. It must be pointed out that unruly conduct does not occur too often, and most of the time roosters behave toward you in the same manner as hens.

Producing your own stock

After your rooster has been with your hens for at least a couple of weeks, you can now start to collect the eggs you wish to hatch and rear.

Storing and sitting eggs for hatching

The process of breeding your own stock is successful only if the birds are healthy and the incubation is done carefully. With many pure breeds the hens will go broody and want to stay in the nest to incubate the eggs, so it is best, prior to sitting the eggs, to collect them every day. You need to remove the eggs from the nest to know when they have been laid so that you can put them all into the incubator or back under the hen at the same time. This means they will all hatch around the same period and not at the rate of one per day for twenty-one days, for example.

It does not matter if the eggs get cold during storage, which can be for a period of up to two weeks, although subjecting them to freezing temperatures does not help. There are many theories on the best way to store eggs, but storing them in an egg box or on an egg tray with the point down is accepted as safe. Once you have collected enough eggs—fifteen is about the maximum you would put under a single hen—take

them out of the egg box or off the egg tray and put them in the incubator or back under the hen. Some poultry keepers seek to replicate what the hen would do, and lay the eggs on their sides. If you do this, you will need to turn them gently at least once a day, otherwise the yolk with the germ attached may rise and stick to the side of the shell. This will result in a deformed chick because, even if the germ develops, the chick will not have had the chance to fully develop in the same way that it would have if the germ was in the middle of the egg.

You will need to select a quiet place for the broody hen to sit for the three weeks it takes for the chicks to develop and hatch. Leaving the broody to incubate her eggs in the place where she has previously laid could result in her being disturbed by other hens wanting to lay in that same nest. The eggs may thus either get cold, or worse still, broken.

Instead, make a separate nest for the broody. The nest should be about the size of the one she usually lays in. Place a turf of grass in a box with the center pushed down and a thin layer of hay on top. It will help if you shape the bottom a little so that the center is lower, to stop the eggs rolling into the corners. Some poultry keepers choose to leave the nest open for the bird to come and go as it wishes. Others close the nest and pick the bird off once a day to let it feed and drink and to empty its bowels. Both methods can be equally successful but unless you can see evidence that the broody has left the nest, you must lift her off as she may start to starve herself and become unhealthy. In the worst cases this may result in her abandoning the nest. If you lift the broody off the nest it is best to check after fifteen minutes if she has gone back on, as sometimes they will just sit down next to it. Do check that no eggs have been broken—if they have, remove the residue and wipe down the other eggs with a warm damp cloth. Do not wash and scrub them. For the last day or two before the eggs hatch do not try to force the hen to leave the nest as she can hear the chicks moving inside the eggs and will wish to stay with them.

Feeding broodies

It is generally acknowledged that the feed given to broodies during the incubation period should be basic and simple. Mixed or straight grains are preferable to layer's pellets. While broody, the hen does not want to be pushed back into lay, and feeding her layer's pellets would cause her to cease being broody, come back into lay, and leave the nest. Also, grains tend to produce more solid droppings, which are cleaner and easier to dispose of. Keep the area in and around the nest clean and free from droppings, otherwise the broody may stand in the mess and foul her eggs when she returns to the nest.

Hatching and rearing

Around the twenty-one-day period the hen will stay on her eggs, although as they start to hatch you will see her raise herself up a little. Even though you may be curious to know how the hatch is going, it is best to leave her to it. If you try to lift her she may stand on an egg that is hatching and crush the chick before it can break free from it. Most broodies will stay on the nest until all the eggs have hatched. Do not worry about feeding her on the nest as she may foul it and her chicks may get stuck in the resulting mess. Once she does leave the nest, check that it is clean if you are leaving her to rear her chicks in that area. If the front of the nest is more than 1 in. high, then place a piece of wood or a brick next to it to make a ramp for the chicks to follow the hen back in. Chicks do not mind falling out, but for the first few days they will find it hard to jump up again.

It may be that you decide to move the hen and her chicks to different quarters to rear them. If this is the case, then a change of bedding is called for. Hay is fine during the incubation period but wood shavings will now be more appropriate. Loose hay can get tangled up around the hen's legs and then strangle the chicks. If something is wrong, you will hear the chicks give a distress call in the form of constant cheeping. On hearing this you should investigate at once, as a chick may have been left in the cold or fallen in the drinker or, in extreme cases, may be tangled in the broody's feathers and dragged around by her. The feeding and watering of your baby chicks will then be the same as if you were rearing them artificially. A good time to take the broody away from her chicks is when they are feathered, although she will let you know when she has had enough of them, usually when she starts to lay again. As stated previously, laying will have ceased during the whole of the broody period.

Stopping broodiness

Unless you are hatching eggs you won't want your hens to remain broody. If you wish to break a broody from sitting, then you should do it before she gets too settled. This should be within the first two or three days of finding her staying on the nest. Keeping her out of the nest does not always work as she may just find somewhere else to sit. If this is the case put her somewhere that is not too comfortable and where she cannot generate any heat—somewhere with a wire or slatted floor is ideal. She should always have plenty of food and water available to get her back into lay as soon as possible. It should only take a couple of days before she wants to rejoin the other hens and stops the clucking noise all broodies make. Unless your hens have a rooster with them and the broody hen has been on the newly laid eggs for more than a day, there should be no visible signs of any germination within the eggs. Collecting eggs regularly during the day and making sure there are none left in the nest for very long will also help deter broodiness.

Pests and how to treat them

For the most part, keeping chickens should be trouble-free, but it would be unfair not to warn you of any of the possible difficulties that may lay ahead. It is likely that one or more of the following ailments will affect your birds at some point in their lives, so it is wise to be familiar with both the ailments and their treatments.

Lice

Although your birds should be free of lice when you collect them, it only takes one or two lice eggs laid in the feathers for them to breed on your birds. There are, however, many effective proprietary treatment powders and sprays available. A good time to carry out an inspection is when your birds are on the perch at night, as it is far easier to pick them up when they are asleep. The main areas where lice live are around the vent, on the thighs, and on the neck. If you find any lice, spray or shake a little powder under the feathers. This is best done outside on a night when it is not too windy, otherwise you may also unwittingly feel the effects of the treatment (in the form of an unpleasant odor!).

If you have a bad infestation you will find clusters of little gray eggs on the feathers around the vent. These are virtually impossible to dislodge unless the area is washed, so repeated application of your treatment, as per the manufacturer's instructions, is necessary to kill the lice as they hatch. To a certain extent, hens can deal with the problem themselves, by "dust bathing," or rolling themselves in dust. This chokes some of the lice and makes them fall off, but a bird can never rid itself completely of lice on its own. Because the eggs are so hard to dislodge, prevention is better than cure so regular weekly inspections should be carried out.

Northern mites

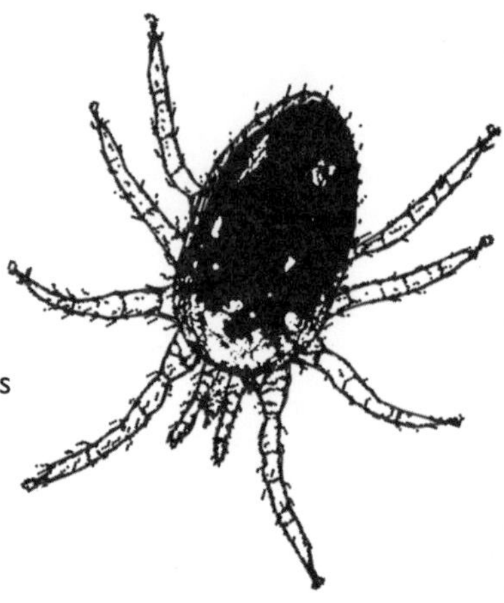

Another little creature that can cause havoc is the northern mite. This lives on the bird and shows itself as damp patches on the feathers, with groups of them crawling around. These patches often occur in the neck feathers and around the vent. Do use a specific treatment for northern mites as others are not as effective.

Red mites

Similar in size to the northern mite but not living on the bird all the time is the red mite. This little creature crawls on to the bird at night and sucks its blood. When not full of blood it looks gray but after a meal it is red. These mites are usually found around the ends of, and under, perches or in crevices; they can also live in nest boxes. Red mites are probably the worst offenders of all because they can kill birds and are especially prevalent in warm weather.

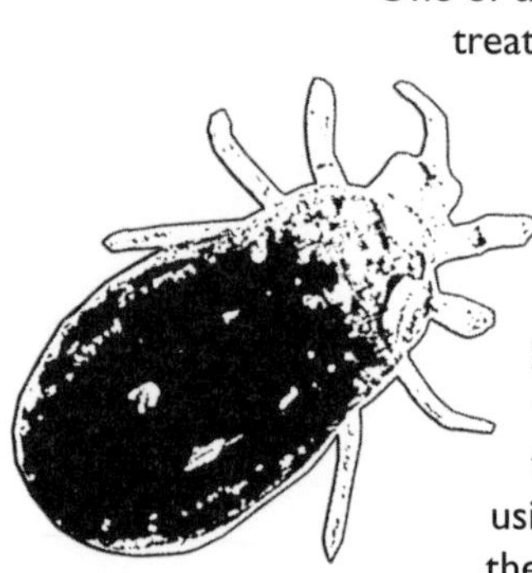

One of the most effective ways of treating an infestation is to treat the hen house itself. The areas where the mites live should be sprayed or powdered after cleaning. They will also live in crevices where boards of the house or the roof coverings join. Spray the house but do pay particular attention to the perch as this is where the bird sits in one position for the longest period. Red mites can live for up to a year without the need to feed off a hen so do take care if housing birds in units that have been used by previous broods. If you are using a broody hen to incubate your eggs do inspect the nest carefully.

It is possible to tell if your birds are infested without even handling them. Listlessness and going off lay are visible signs of infestation. If the house has too many red mites the birds may be reluctant to go in to roost at night. One old-fashioned treatment is to treat the house inside and out with creosote when it is empty and cleaned, but do leave it empty for a few days after treatment, as the vapors can be quite noxious to the birds when closed in at night. An added bonus is that creosote does last longer than many other modern treatments and it also helps to preserve your unit's woodwork.

Scaly leg mites

Another little parasite that could cause you trouble is the scaly leg mite. It is much more visible than other mites because, in the worst cases, it lifts the scales on the legs and leaves a white encrustation, crippling the bird. It takes a very bad infestation to make a bird lame, but unless treated it will eventually do so. One of the best cures is a sulphur ointment as it will not only kill the mite but also leave the scales on the legs smooth and shiny unlike other preparations, which could leave them sore and dry.

Diseases

There are a number of diseases that afflict poultry, but thankfully for the small keeper, they are most prevalent only in the larger commercial units. Because commercial birds are kept in intensive houses where thousands live together, the chance of disease spreading is increased. You could compare it to the way children catch colds from one another when in close proximity in a warm classroom. Commercially farmed birds have often been vaccinated against diseases for many generations so they no longer have a natural immune system. Small free-range flocks do not come into contact with these larger flocks, and so from one generation to another have developed a fair immunity to diseases.

Avian Influenza

There is no better illustration of this immune system in action than the scare about Avian Influenza (AI), or bird flu. Many millions of chickens were predicted to die from a mutation of AI and human influenza. In most cases, since it came to countries outside Asia (where it originated), it has been confined to wild waterfowl, especially swans, which have proved rather susceptible, plus large intensive collections of poultry. The birds of small poultry keepers in the vicinity of outbreaks have been tested but have in virtually every case been found to be free from the disease.

Far from the original ideas that were put around, poultry do not all have to be kept inside under cover. Putting them under cover is only necessary if there is an outbreak of AI in the immediate area where you live. Some governments have brought in a statutory registration system for all small poultry keepers with more than fifty birds. This shows that they do not see the threat of the disease spreading through small flocks. If you do register and an outbreak occurs, the government will immediately communicate with you by text or e-mail. The problem has far more to do with media overreaction, than the actual threat from the disease. However, vigilance is necessary, and should you get sick birds with flu like symptoms, such as sneezing, runny eyes, listlessness, and more importantly high mortality, you should seek immediate advice from your veterinarian.

Mycoplasma

A disease now often called mycoplasma was for many years known as roup, or the common cold. In former times, when birds used to be kept outside rather than in the intensive, indoor systems used today, the only treatment for a severe outbreak of roup was culling. There were none of the modern medicines back then, and so one had to cull those birds that had the roup, hoping that the others in the flock would survive. It was, however, known for individual birds to recover if their nostrils and throat received an application of a mentholated topical ointment used on humans and available from any pharmacist.

Mycoplasma occurs most frequently these days in intensive situations where a period of stress has rendered one or several birds vulnerable. The disease then spreads quickly to the rest of the birds.

The most obvious evidence of mycoplasma is a discharge from the nostrils that is identifiable by a particular soft, sweet smell. The recognized treatment for small flocks is Tylan Soluble or Baytril. If there are a number of cases, then clean out the house and disinfect it with a recommended disinfectant such as Virkon. Some veterinarians recommend a course of antibiotics but this can be very expensive and, on the whole, will not cure the problem.

Mareks disease

Mareks disease is most prevalent among commercial flocks, although certain breeds of chicken seem to be more susceptible to it. It is more common in intensive conditions because it is spread in dust, including that from the feather follicle shed from newly grown feathers. It normally occurs in growing birds after six weeks, usually between twelve and sixteen weeks, and because it affects mainly the female of the species, it manifests itself when there is a hormonal change as they approach lay. The signs are very obvious in that the wings and legs show a paralysis and the birds look really ill. Once they have reached this stage there is no cure and you should kill them immediately. There is, however, a vaccination, but this is only effective if administered at one-day-old, or at least before three weeks. The two breeds most prone to Mareks disease are Silkies and Sebrights, and for this reason most are vaccinated. The disadvantage of vaccination is that these breeds' and commercial flocks' immunity to the disease has reduced over the generations.

Coccidiosis

Coccidiosis is a disease usually caused by damp, dirty conditions, and it affects baby chicks' guts and causes internal bleeding. Although they will hopefully recover from it, it can slow their growth and they may never completely recover. Most crumbs and mash feeds contain ACS (anti-coccidiostat), an additive that protects the chicks against infection. Symptoms of coccidiosis include listlessness, droopiness, and red droppings. Since this disease has a forty-eight-hour infection cycle, it is essential that you change the litter of young chicks very regularly and clean up any soil or spills right away, putting down fresh litter, particularly if you see any red droppings, because the bright color will attract the chicks to peck at them.

All the above-mentioned diseases are the only ones commonly seen in poultry. They are not that prevalent, and free-range, pure-bred stock has a much better chance of escaping infection. Although clinically clean conditions are often recommended, a good routine is to make sure that your birds live in clean, dry, and well-ventilated conditions.

Less common problems

There are a number of other, less common troubles that can affect the well-being of your hens. These should be dealt with sooner rather than later to ensure a happy, healthy brood.

Egg eating

When you bought your hens you probably thought that eating the eggs would be enjoyed only by you and your family. Unfortunately, and usually by accident, egg eating is a vice that can occur among the hens themselves. It tends to happen when two hens try to share the same nest box to lay and an egg gets broken. Bright colors attract the birds and the yolk looks particularly appealing. If the egg shells are strong enough this problem should not arise and one broken egg should not set off a lifetime of bad habits. You may, on closer observation, find that only one hen is eating eggs. If this is the case, you should isolate her.

If you find that a hen is eating her own eggs, then there are various cures. Some say that adding mustard or curry powder to a broken egg's content deters the bird from eating it. Others say using a pot egg (a clay egg) or even a golf ball will make the hen think its efforts are not worthwhile. If the nest boxes are really open and well lit, fix some strips of cloth or sacking across the front. The hens will still know where they normally lay, as they are creatures of habit and will still go in, but will not be able to see the eggs.

Bleeding

Sometimes through overcrowding, bullying, or the attention of a rooster during the mating process, when the male's toenails can damage the back of the female as he holds on to her, external bleeding can occur. This bleeding can lead to feather pecking that, in extreme cases, can result in cannibalism. If you can, spot this problem early and isolate the affected hen until the bleeding dries up. The problem here is that the feathers may have been plucked out, and when young feathers grow over the damaged area they are very soft and can easily be damaged and bleed again.

Should this happen, spray the affected area with an antiseptic spray or use the old-fashioned remedy of Stockholm tar. This thick and sticky antiseptic is best applied with a wooden spatula and it is very effective, as the hen will find the area difficult to peck at. As pecking usually develops through boredom, try hanging up a cabbage stump to give them something soft and juicy to peck on. Polystyrene has also been found to occupy hens' attention. Do not give them small pieces, but rather place a large block of it in the house. It is not understood why hens find it so attractive to pick at. They will swallow it, so do not let the birds gorge themselves on it or it may affect their digestion; just use it in times of trouble. In small amounts, polystyrene does not seem to do the birds any damage.

A bleeding comb is not a serious injury as it will, in 99 percent of cases, heal naturally.

Bullying

Bullying is not usually an issue if your hens have all been brought up together or brought into the hen run at the same time. If, however, you wish to introduce new hens, then there are a number of tips you should follow. It should be noted that whatever you do there will be some fighting if you bring in new stock. The common phrase "establishing the pecking order" springs to mind. When you watch your hens at the feed trough you will note that the same bird is always dominant. Once the pecking order has been established it will usually last forever, so any new hens on the block will need to deal with this.

If possible, it is recommended you put both new and established birds in a different house to the one the established birds have been living in. This is so that none of the hens will feel completely secure, as both new and established birds are in an unfamiliar environment. It is best to introduce them at night and then the next day scatter food around the run and house so that if any bullying does occur they all have a chance of getting some food. Whatever you do, unless a hen is really being pulled to pieces, do not take her out as she will suffer even more when you reintroduce her into the house. She will be the main object of attention again and will have residual feelings of inferiority.

Crop impaction

Unless your hens really eat down short any grassy areas, it is best to periodically mow the grass. When grass grows long it is usually tough and can cause crop impaction. Once the grass has passed into the gizzard there should be no problem but long, tough pieces of grass can hold up the flow of food, leading to a collection of sour, fermenting food in the crop. You will usually find that the bird is not eating and the crop area appears rather large and firmer than usual.

If this happens, isolate the bird and put it on a water-only diet for a couple of days. You can help by massaging the crop, and if the food does not move, turn the bird upside down and squeeze the mess gently out of its mouth. String used to tie bales of hay or straw that has accidentally been swallowed can also cause similar problems. Unless it comes away easily, if string gets down the bird's throat, simply cut off the visible piece and hope that only a small piece will have to pass through the bird.

Finding a vet

A great proportion of veterinarians will admit they know little, if anything, about poultry. It is a good idea, therefore, to make a few inquiries to check whether there is a poultry-friendly veterinary practice in your area. Many birds have died as a result of inaccurate diagnoses and incorrect and often expensive treatments. These days many pet stores will have a qualified member on staff who can offer advice or check with someone who will suggest the proper treatment. In these days of Internet communication, you may be able to find a site or even a local or national poultry club where free advice is available from those with firsthand experience.

Chickens and other household pets

If you already have pets, it's important to keep them under control when they are around chickens. Most domestic pets will initially find chickens intriguing but after a while will lose interest and accept them as just another addition to the household. Dogs can be a challenge, and if your hens roam a large area they may see them as something to chase. This will stress your chickens and they may consequently cease laying. In extreme cases, it has been known for dogs to catch and kill hens. If this happens, make the dog sit, hold the dead bird in front of it, and tell it in your best stern voice that it has done wrong. It will very soon realize that your tone of voice and the dead bird are related, and with a little extra vigilance it should not happen again.

Cats may challenge hens but in most cases it either results in a standoff, or one or the other accepts defeat, without any injury but with slightly dented pride. However, chicks are a different matter, and unless they are accompanied by a broody hen, in the eyes of a cat will look similar to a wild bird. It is a very brave cat that will challenge an offended broody hen!

You may already have horses or other large animals, and if this is the case the birds should not be harmed unless they are stepped upon. Your main concern is to avoid damage to the house and run. The best, and easiest, way, especially if your house and run are movable, is to surround them with electric fencing; a single strand should be sufficient. Heaps of horse droppings can kill the grass underneath them so if left out to range with horses or other animals, hens can have a beneficial effect by scratching at the heaps on the lookout for grains, thus helping to spread the droppings.

Some breeds of dog can be trained to protect hens against predators. These breeds are happy to spend a lot of their lives outdoors and often have been bred to look after livestock. Even if they do not drive predators away they will still raise the alarm and alert you to any potential danger in the chicken coop. If the dog is left alone with your hens, do make sure it doesn't eat all the eggs; this will often happen when the dog attempts to pick up an egg, so you will have to train it not to do so. In many cases the egg will not break but will just be carried away and buried. If, however, an egg does break in its mouth, the dog will eat it, along with all the others!

Taking a vacation

This refers to you and not your birds! Where livestock is concerned it has often been said that no one looks after them quite like the owner, so it sometimes feels impossible to get away. If this happens, the birds become a chore. Going away for a day should be quite easy, even if you have to get up a little earlier to let the hens out and feed and water them. If you are going to be back after dark, you have the option of either giving them a larger feed in the morning and trusting all will be well until you return, or getting a neighbor just to close the house for you when it gets dark.

Things aren't quite so simple if you are taking a longer vacation. Planning should take much longer to make sure that any last-minute problems do not make your time away worrisome. The best time to go away is in the late summer or early autumn, when the days are still fairly long but all your hens are well established in their quarters. Never leave soon after you have bought new hens or moved them to new quarters. Equally, leaving your birds in the depth of winter when snow is on the ground and everything is frozen over is not going to go down too well with your chosen birdsitter.

Who you choose to look after your birds in your absence requires some thought. If you only have a few hens, then a neighbor may be able to help. Do make sure they are reliable though, and have both the time and dedication to be available every day. If they cannot make themselves available for one of the days, try to get another friend or neighbor on board, too. It is always a good idea to write down a list of chores that need to be carried out, and if you have more than one carer, creating a daily rotation will save any unfortunate mishaps or costly mistakes.

A fellow poultry keeper could help, so long as they do not have too far to travel. Children can also be very good, so long as their parents agree and know what is expected of them. Or you may decide to have a house sitter, and in the majority of cases this will be the best solution. Do try to impress on anyone you ask to do the work not to change the way you do it. Hens are creatures of habit, and moving feed and water troughs to other areas of the run is not always advisable. If possible, try to get your replacement in for a day before you leave so that you can show them the ropes. Paying your helpers for their work, even if it is in eggs, is always appreciated.

Do not leave any tasks that you intended to carry out unfinished. Always mend those rickety gates or fences so that they do not fall apart the

minute your carer touches them. If you have water barrels for refilling water founts, make sure they are full. It is best to buy more feed than you would actually use as the birds will often be slightly overfed. You should, however, make sure your carer knows how much feed to give them. If you are lucky enough to have a chicken expert nearby, do ask if you can pass on their phone number to your carer, who will always appreciate a second opinion should it be necessary.

Once you have ensured all is in order and you have left for your vacation, stop worrying about what is happening at home! There is nothing you can do and what is a vacation for but getting away from things? When you return, do not feel too upset if everything is not as you left it. After a few days it will all be forgotten and you will be back in your old routine, enjoying your chickens.

Exhibiting your birds

After a few years' experience keeping your hens and perhaps breeding some of your own stock, you may be so hooked on everything to do with keeping chickens that you wish to enter the competitive arena. Even if this is not your scene, it is always a fun day out to visit rural or agricultural shows where poultry are being exhibited. You will see lots of rare and spectacular breeds, pick up advice and tips, and be able to buy all sorts of new kits and equipment. Poultry keeping is enjoying a great resurgence in popularity and there is a whole network of experts and enthusiasts that you can tap into.

Local and national clubs

The fact that there are other people out there with a similar interest as you and who are available to offer help and advice is a comforting thought to many novice poultry keepers. Until recently, information on these organizations was hard to come by, but thanks to modern communications just about all of these clubs now have a Web site and are available either by e-mail or phone. There are also national bodies worldwide, often with a designated committee whose role it is to offer help and advice to the newcomer. These bodies offer information on clubs for specific breeds of chickens, turkeys, ducks, and geese. Additionally, there are clubs that cater to small poultry keepers. Poultry forums are a great way to exchange news and views but don't always expect expert advice here; these Web sites are mainly for novices feeling their way around and sharing their personal experiences.

Local clubs are usually the most useful for the novice. They will often hold shows of pure-bred birds but will also have evening meetings and gatherings at members' homes. These events are informal and usually include refreshments, a talk or demonstration, which will often include audience participation, followed by a question-and-answer session. Although there may be a number of experts there, many will admit they are still learning.

Poultry shows

Poultry shows are a great way of seeing the myriad of colors and the many shapes and sizes of breeds. The advantage of attending these shows is that you get to see firsthand both good and bad examples of the particular breeds you are interested in, often with live demonstrations. Small local shows will usually last a day, with birds being brought to the show early in the morning. All show pens are numbered and the exhibitor is allocated certain pen numbers for each of their birds. The classes for them usually follow a breed and if the breed is popular it is then subdivided into colors. There are classes for large fowl and bantams, so do make sure if entering the show that your bird goes in the correct class. Trying to fit a large fowl into a bantam pen looks bad and is inhumane. If you have cross-bred or hybrid hens you can still enter their eggs in the egg classes, and sometimes a show will have classes for the likeliest layer.

Shows will engage qualified judges though show etiquette means that if you have questions regarding why the judge either loved or hated your bird, you should not engage them in conversation until the judging of all the birds is over, otherwise it might be interpreted as you trying to influence their decision. Like all branches of poultry keeping, showing can be addictive and you could find yourself traveling all over the country and even the world! It has been known to open many doors.

In most countries the major shows are run by or combined with meetings of the parent body governing all matters relating to small poultry keepers. These bodies will be engaged with government bodies to help minimize the effect of legislation on small poultry keepers. Many items are debated by these bodies during the year and they will also work with similar organizations in other countries, sharing their knowledge and concerns.

The parent bodies are also responsible for producing books, which are regularly updated and contain the Standard of Perfection for each of the known breeds of poultry in that country. These books contain the history of breeds, shapes, sizes, and color patterns. The breeds are sometimes of local origin but many have been imported from across the world. Whichever country the breeds are from, they will have been bred by individuals originally and then taken up by a group of poultry fanciers. These will often form a club to foster the interests of these birds. After they have been bred to show, a large percentage will reproduce the parents' characteristics. The new breed will then be submitted for approval, eventually gaining entry into the book of standards. This book is used by breeders and judges alike and provides invaluable information on what a bird should conform to. It is generally accepted that type (shape and carriage) makes the breed and color the variety. An old saying is "You build the house first, then you paint it!"

Poultry breeds

Chickens do not all look the same! Legs, tails, carriage, and feather color and texture vary between the many different breeds. The following pages showcase just some of the myriads of breeds available to help you select one that suits you and your needs.

The available choices

Now that you are well-versed in the ways of keeping chickens, you will need to decide which breed you want to keep. There are many choices available: pure breeds, which are divided into heavy and light breeds; bantams, which are also divided in the same way with the addition of true bantams; and finally hybrids.

Not all breeds are equally popular so it is a good idea to check if the one you like is available without having to travel too far. Most heavy breeds lay eggs with a colored shell, and light breeds, which have a Mediterranean influence, lay white-shelled eggs.

So, where do we start? Most readers new to keeping chickens don't realize just how many different breeds there are, each with their own very distinctive physical characteristics, laying habits, and personality!

Heavy breeds

Heavy breeds are ones that are heavier in body weight than light breeds. They are less prone to flying and so are usually more tame than the light breeds. Because they are slightly larger, they tend to eat a little more than light breeds and will require slightly more space.

Marans

Key facts

- **Cock:** 8 lbs.; **Hen:** 7 lbs.
- **Available in:** large fowl and bantam
- **Preferred climate:** virtually any

This breed has been in existence for a long time and was originally produced as a fast-growing, white-skinned-meat bird for the Paris markets. It was only when it was found to lay dark brown-shelled eggs that its popularity really took off. The original Standard called for birds with a cuckoo colored plumage and white legs with no feathers. Latterly new colors have been developed in France that are much more colorful with the added bonus that they lay even darker-shelled eggs. These recently developed additions have feathered legs, which sets them apart from the original Marans.

Because Marans are a heavy breed they seem to thrive on even the wettest of soils. They are prone to a certain amount of broodiness although the "newer" varieties are less so. Of a relatively tame nature, these hens make good pets. The male birds of some strains can be somewhat vicious, but most people will keep only the females for eggs so this is not usually a problem. Because they were originated for the table they mature relatively quickly.

Rhode Island Red

Key facts

- **Cock:** 8.5 lbs.; **Hen:** 6 lbs. 10 oz.
- **Available in:** large fowl and bantam
- **Preferred climate:** virtually any

This breed's name tells you all about it. Red in color, it was developed in North America to withstand cold winters and lay lots of light brown eggs. In the last century it was one of the breeds that, in its true form or crossed with others, was the backbone of the poultry industry. A pen of Rhodes in the bright sunlight looks very attractive. The male bird is of a deep red color and has a black tail, and the female's colors are the same but without the male bird's sheen. They have a distinctive shape, which is best described as brick shaped or oblong. Bright yellow legs with some brown and red tinges set it off. Rhode Islands are reliable layers of a fair number of eggs and do go broody but not for long periods. They seem to have a quiet nature and thrive in most conditions.

Sussex

Key facts

- **Cock:** 8 lbs. 13 oz.; **Hen:** 7 lbs.
- **Available in:** large fowl and bantam
- **Preferred climate:** all

The Sussex was created in Britain and has a long tradition of originally supplying the London markets. Most assume that light is the only color available but there are a number of others, although these are less popular. The light has a white ground color with black striping in its neck and wings, and a black tail. The legs are white. Another color that has vied with the light for popularity is the speckled. In this you have an attractive mix of mahogany, black, and white, in that order, from stem to tip of feather.

Other less common colors include white, buff, red, silver, and brown. The first three are, however, the ones that are most readily available. They are good layers of tinted or very soft, creamy brown-colored eggs. They can compete with the other popular breeds in terms of numbers of eggs produced per year.

Wyandotte

This is a breed of North American origin and the white variety was at one time at the top of the list of pure breeds for egg laying. Nowadays many of the breed's other color varieties can compete on that feature while often being a more visually attractive bird. When looking at the comb you will notice that it is not the same as that of other chickens. Its formation is spread across the top of the head rather than standing up in a single row. This type of comb is called a rosecomb and it is a feature in a number of other breeds. As it lies close to the head it is not as liable to damage.

As mentioned, the white variety was best for laying but unless a utility egg laying strain can be sourced you will find the heavily feathered versions can be very poor layers. Very popular are the silver laced and blacks. Other colors seen are gold laced, blue laced, and partridge.

Key facts

Cock: 8 lbs. 13 oz.; **Hen:** 7 lbs.

Available in: large fowl and bantam

Preferred climate: cold to moderate generally, although some varieties have been adapted to warmer climates

Barnevelder

This Dutch breed was originally developed to supply the egg markets in the town of Barneveld. It is an attractive breed that has one very dominant color variety: the double laced.

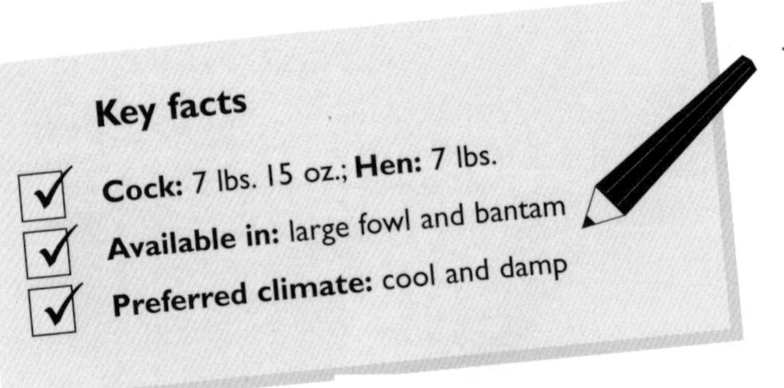

This has a lovely mahogany ground color with black lacing both around and in the feathers. The breed is known for "beauty with utility" and it is certainly true.

If you are looking for a dark brown egg layer (not as dark as the Marans) that looks good and is very active, you should consider the Barney. Other color varieties are black and blue laced.

Orpington

These are well known through British Royal patronage of the buff variety. It is a British breed created by William Cook in Orpington, Kent. It has developed more into a show breed but will still perform well in all sorts of conditions. The buff is most popular and is a golden buff color. The black, blue, and white varieties tend to be more fluffy than the buff but are equally attractive. They lay fair-sized tinted colored eggs.

Key facts

- **Cock:** 9 lbs. 15 oz.; **Hen:** 7 lbs. 15 oz.
- **Available in:** large fowl and bantam
- **Preferred climate:** cool rather than hot

Australorp

This breed should be closer to the original Orpington in looks and character as it was re-imported from Australia as a breed that, although useful for showing, has had its laying capabilities maintained. It is kept in the largest percentage as a black although efforts have been made to popularize the blue and white. A very smart bird similar in shape and plumage to the Barnevelder, it is capable of laying plenty of tinted eggs.

Plymouth Rock

This breed originates in North America and for quite some time it was regarded as a very good dual-purpose breed, laying plenty of eggs and growing quickly. Many strains of the buffs laid very well, and up until recently the white was used extensively in producing table chickens. The other color that is seen in any numbers is the barred. In this color, as with the cuckoo coloring, the feather is marked across with alternate black and white markings. All three of these colors are still useful and well worth looking at.

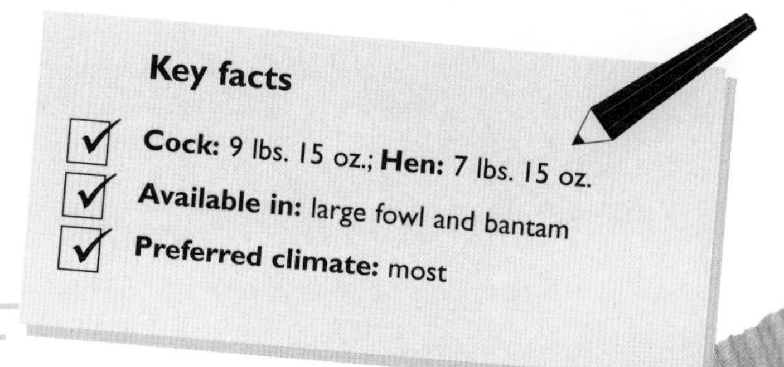

New Hampshire Red

Key facts

- **Cock:** 8 lbs. 6 oz.; **Hen:** 6 lbs. 10 oz.
- **Available in:** large fowl and bantam
- **Preferred climate:** cool

This breed was produced in the state of New Hampshire in the early 1900s and was accepted as a pure breed in 1935. The females had to be able to withstand the very harsh winters while still laying plenty of eggs, and the males to produce good bodies, carrying plenty of meat. It was therefore a useful breed in the creation of the first broiler chickens. The New Hampshire Red's coloring is much more orange than that of the Rhode and the male has light yellow-orange hackle feathers. They are not so common so if you find them they are worth the investment.

Dorking

You may, in your search for something different, come across the Dorking. This is a breed with a long history. A fowl very similar to the Dorking was described by a Roman writer during the conquest of Britain. The name stems from the fact that fowl such as these have been found around Dorking in Surrey. They were bred for the table and have a long, low body and a stately walk. Another feature seen in this and other table breeds is the five-toed foot: three at the front and two (one above the other) at the back. They come in a variety of comb shapes and five colors: silver-gray, dark or colored, red (red partridge), cuckoo, and white. This breed appreciates outside runs, so long as they have plenty of shade and protection. Despite having fairly long wings they do not fly much. They lay a fair amount of creamy tinted eggs.

Key facts

Cock: 13 lbs. 14 oz.; **Hen:** 9 lbs. 15 oz.
Available in: large fowl and bantam
Preferred climate: moderate

Faverolles

Key facts

- **Cock:** 9 lbs. 15 oz.; **Hen:** 8 lbs. 6 oz.
- **Available in:** large fowl and bantam
- **Preferred climate:** cool

This breed was produced in France for the table bird market and has a big, wide body, described as "cloddy." Besides having five toes and feathers down the outside of its legs, it also has head adornments in the form of a beard and side whiskers. It is of a quiet nature and does well out on open runs. It is a good layer of tinted eggs of good size. It comes in a number of colors but the original color, salmon, is the most popular. The male has a creamy colored hackle with a black beard, chest, tail and thighs, and rich cherry colored shoulders. The hen has a white beard, body, and thighs, with the rest a light salmon color. The other colors available are black, white, buff, blue, cuckoo, and ermine.

Croad Langshan

Originally from Vietnam, this breed takes its name from the person who first imported them into England, Major Croad. There are many variations of this breed, including the Black, Modern, German, and American Langshan. The Black is the original and most popular. In all types the bird is tall with a short back; the Croad and American both have a high tail carriage. All variations have a light fringe of feathers down the outside of the legs, except the German, which is clean. A feature of the Croad is that its brown eggs have a plum-like bloom on them. Other color varieties in each type are blue and white. It is a good layer and if cockerels are reared for the table their flesh is of a good texture. It is a breed that thrives in small open runs. If kept in houses with connecting runs these need to be high to allow for the height of the bird.

Key facts

Cock: 8 lbs. 13 oz.; **Hen:** 7 lbs.

Available in: large fowl and bantam

Preferred climate: fairly adaptable, even to hotter climates

Frizzle

This is a distinct breed, and although it looks like an ordinary hen in shape, it has feathers that curve outwards and upwards instead of lying smoothly along the bird's body. Despite appearances it is hardy and can tolerate wet and cold weather. It has a reputation for being broody and is a good layer. If breeding Frizzles, not all will have frizzle feathers but if a non-frizzle is bred back to a frizzle it has a good chance of being frizzle, and the feathers, which can be narrow, should be broader. The main colors are white, black, blue, and buff. It lays creamy white eggs.

Key facts

- **Cock:** 7 lbs. 15 oz.; **Hen:** 6 lbs.
- **Available in:** large fowl and bantam
- **Preferred climate:** moderate

Cochin

Key facts

- **Cock:** 12 lbs. 12 oz.; **Hen:** 11 lbs.
- **Available in:** large fowl and bantam
- **Preferred climate:** all

The Cochin is reputed to have started the boom in keeping pure breeds back in the 1850s. It was a craze that saw these birds imported from China, fetching what was then a small fortune. They revolutionized the way people thought about poultry keeping, which had been, until then, primarily about meat, with egg production taking a back seat. Although not as good for the table, these birds produced eggs throughout the winter and would lay around 150 eggs per year—an unheard of figure in those days. It was, however, their novelty that led to their downfall: high prices could not be maintained and the fashions changed. The breed became one for the show pen and the amount of feather was increased by selective breeding. Selective breeding led to lowered egg production and although Cochins can still lay a fair number you must look to them for their beauty rather than the quantity of eggs they lay. It is a breed noted for its feathers and curves, and its body is covered in a long, soft plumage extending down the legs and to the end of the toes. It is a quiet bird and a very reliable sitter. Six colors are available: white, black, blue, buff, cuckoo, and partridge. They lay tinted eggs of average size. A claim to fame is that many of the brown egg-laying breeds have the Cochin somewhere in their original makeup.

Brahma

Key facts

- **Cock:** 11 lbs. 14 oz.; **Hen:** 8 lbs. 13 oz.
- **Available in:** large fowl and bantam
- **Preferred climate:** most

This breed became popular on the back of the Cochin. While not attracting the same exorbitant sums as the Cochin, the Brahma still did well. Its origin is open to debate; some say it hails from the Brahma Pootra river in India; others that it was created by a great publicist called Burnham in North America who made claims of its wondrous appearance, gentle nature, and egg-laying qualities. To the uninitiated, this breed is of a similar appearance to the Cochin, but it does have certain differences. Its head has a pea comb (three small parallel rows of small serrations) rather than a single comb.

It is also more closely feathered and longer on the leg than the Cochin. It comes in many colors, most of which are of very recent creation. The two original and still most popular colors are the light and the dark. The latter is of a silver-gray color, with the female having a light lacing around the feathers. Other colors are buff Colombian, gold partridge, blue partridge, and white.

Rare heavy breeds

A number of other rarer, heavier breeds of chicken are kept in relatively small numbers. They can be just as productive as their less rare counterparts, so do look into these. There are a number of other breeds, too, but these are not readily available, so unless your agenda is to rescue a very rare breed, stay with those on this list.

Houdans were once very popular as a white egg-laying table breed. Of French creation, they come in just one color: black with white mottling. They have five toes, a large crest of feathers, and also a beard and muffling.
Cock: 7 lbs. 15 oz.; **Hen:** 7 lbs.

Jersey Giants (right) are the heaviest breed of chicken and lay a good number of large tinted eggs. They are found in black, white, and blue, although only the first is seen to any extent.
Cock: 13 lbs. 14 oz.; **Hen:** 9 lbs. 15 oz.

Indian Game, **Malay Game,** and **Ixworths** are all primarily table birds and do not warrant keeping if your main aim is to produce eggs. They do, however, carry a lot of meat on their bodies and although they grow more slowly than broilers, they produce a very tasty meal.
Indian Game: **Cock:** 7 lbs. 15 oz.; **Hen:** 5 lbs. 15 oz.
Malay Game: **Cock:** 11 lbs.; **Hen:** 8 lbs. 13 oz.
Ixworth: **Cock:** 8 lbs. 13 oz.; **Hen:** 7 lbs.

Orloffs are a Russian breed, having a rather cruel face with a flat comb covered in short, hard, spiky feathers, along with a small beard and whiskers. It lays a fair number of creamy tinted eggs and is found in spangled, mahogany, white, black, and cuckoo colors, although the spangled is mostly seen.
Cock: 7 lbs. 15 oz.; **Hen:** 5 lbs. 5 oz.

Light breeds

Virtually all light breeds are of northern European and mediterranean origin. They are generally of a more nervous and therefore more flighty nature but should lay more eggs and consume less food than the heavy breeds.

Leghorn

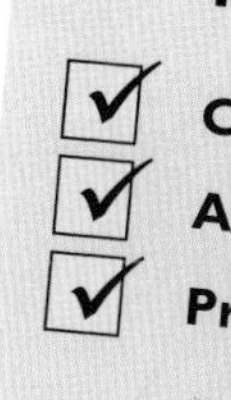

Key facts

Cock: 7 lbs. 8 oz.; **Hen:** 5 lbs. 8 oz.
Available in: large fowl and bantam
Preferred climate: moderate to warm

If you wanted plenty of eggs from a bird that does not go broody and has a small appetite, then this is the breed of choice before the advent of hybrids. British breeders developed a strain of white Leghorn that was exported all over the world and especially so into North America. Leghorns were imported from the Mediterranean and are part of a group of breeds that lay white eggs. There are still some utility strains that are bred for egg production. Leghorns are able to fly over low fences so make sure any fences are at least 6 ft. 5 in. high. Males have a large comb that can be damaged by frost so protect it in severe weather; an application of petroleum jelly will help. Unlike heavy breeds, this and other Mediterraneans have white earlobes, so do not think your bird is unwell. There are a number of other color varieties, including black, blue, brown, cuckoo, and exchequer. The latter has an unusual color pattern where the base color is white with irregular-shaped black mottling scattered over the body. This variety lays well.

Ancona

The Ancona may at first be confused with the Leghorn, but it is a separate breed. Its ear lobes are creamy rather than white and it comes in one distinct color: a black ground color with each feather distinctly tipped with white. The feathers are rounded at the tips and in the shape of a "V" where they join the black. It is a breed that, although known for its plumage, has always retained its utility properties. It can have a single or rose comb and the legs are yellow with green/black mottling. It thrives in open spaces, although it can also be kept in small runs.

Key facts

- **Cock:** 6 lbs. 10 oz.; **Hen:** 5 lbs. 8 oz.
- **Available in:** large fowl and bantam
- **Preferred climate:** moderate to warm

Andalusian

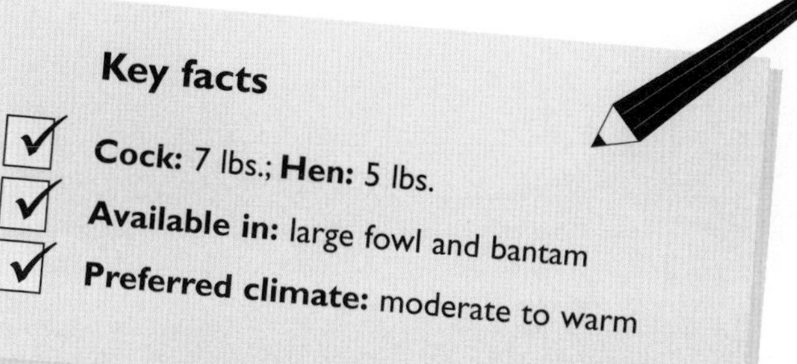

Part of the Mediterranean group, Andalusians come in one color, blue, of a very light shade. If you are breeding or buying hatching eggs you should be aware that blues do not breed true to color; the offspring of blue males to blue females are 50 percent blue, 25 percent black, and 25 percent splash (a silver-white color with splashes of blue). However, breeding blacks and splashes to one another will give you 100 percent blue. You must, however, use a blue-bred black for this to happen. This is known as the Mendelian inheritance law. Andalusians lay fair-sized white eggs. It has a single comb and blue legs.

Minorca

This is the largest of the Mediterranean breeds and lays the largest eggs. It is usually found in black, although a few whites and blues are around. The black Minorca looks very smart and has a bright red comb and very large almond-shaped white earlobes. Despite being bred for exhibition purposes it still retains its laying capabilities and is fairly popular. It is a breed that is happiest on open runs, although it does still produce plenty of eggs when kept under cover.

Key facts

- **Cock:** 7 lbs. 15 oz.; **Hen:** 7 lbs.
- **Available in:** large fowl and bantam
- **Preferred climate:** moderate to warm

Spanish

This is somewhat like the Minorca but its full name of White Faced Black Spanish gives a clue as to its appearance. The comb and wattles are red but the rest of the face is covered in a soft white flesh similar to some breeds' earlobes. It is a breed that has a limited appeal, although it has been around longer than any of the other Mediterranean breeds.

Hamburg

Key facts

- **Cock:** 5 lbs.; **Hen:** 4 lbs.
- **Available in:** large fowl and bantam
- **Preferred climate:** cold to moderate

Despite the name, Hamburgs are reputed to be of English origin, having been found long ago in the north of the country where they were known under various names, including Moonies. Long before imported breeds were used for egg laying, this breed produced enough eggs to earn itself a reputation as a reliable layer. It is a light and rather flighty breed but it has a very attractive plumage. The most common color is the silver spangled, with a white ground color and each feather finishing in a perfectly round black spangle. There is also a gold variety that has a rich bay ground color and a mainly black tail. There are gold and silver varieties of pencilled, too, where each feather has a distinct fine black pencilling across the feathers which are quite tightly packed There is also a black variety. The breed lays plenty of white eggs that are a good size. It can be kept in confinement but as with many other breeds it does better outside.

Old English Pheasant Fowl

Very similar at first glance to the gold spangled Hamburg, this is a breed from the North of England. Instead of round spangles it has crescent shaped black markings on the ends of the feathers. Some examples of a silver variety have been seen but they are very rare. Similar in many aspects to the Hamburg.

Key facts

Cock: 7 lbs.; **Hen:** 6 lbs.

Available in: large fowl only

Preferred climate: cold to moderate

Redcap

This is sometimes known as the Derbyshire Redcap as it has always been found in great numbers in the county of Derbyshire. A very workman-like bird, it does not take well to confinement, preferring to roam free on the rugged Derbyshire hills. It gets its name from the fact that it has a very large rose comb, which looks like a cap positioned on its head. Although its head-points may seem extreme in that its comb can be affected by cold or frosty weather, it is a hardy bird and is more a utility than show bird. The coloring is darker than the Pheasant Fowl and the eggs are white.

Key facts

- **Cock:** 6 lbs. 10 oz.; **Hen:** 5.5 lbs.
- **Available in:** large fowl only
- **Preferred climate:** cold to moderate

Scots Dumpy

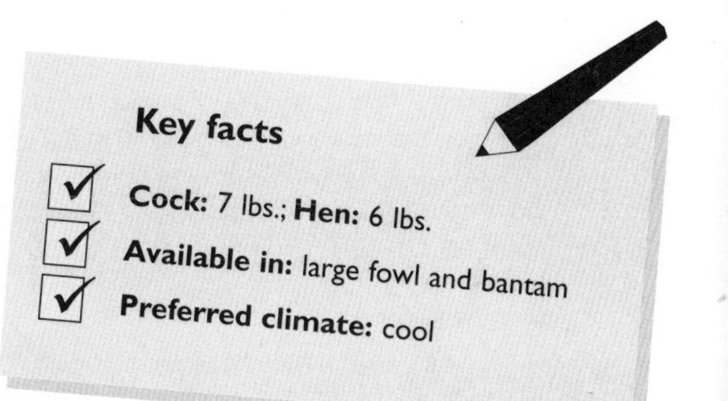

This quaint breed has a long history. It was originally bred in Scotland and is therefore able to withstand the rigors of the winter weather. It gets its name from the shortness of its legs that, when a good specimen is on the move, causes it to waddle like a duck.

It is said that on the islands off the west coast of Scotland these birds could be seen out in a gale without being blown over! Despite its short legs the body is long and flat, and the males carry a surprising amount of flesh. Dumpies lay a fair amount of tinted eggs and make very good, careful broodies. The most common color is cuckoo, although just about any color is accepted. It is more popular now than it has been for many years.

Scots Gray

As the name suggests, this uncommon breed comes from Scotland. It has long legs and a curved body with a long neck and tail. It comes in one color: a steely gray ground color with evenly spaced black barring. Although rare, it is hardy and will lay a good number of white eggs.

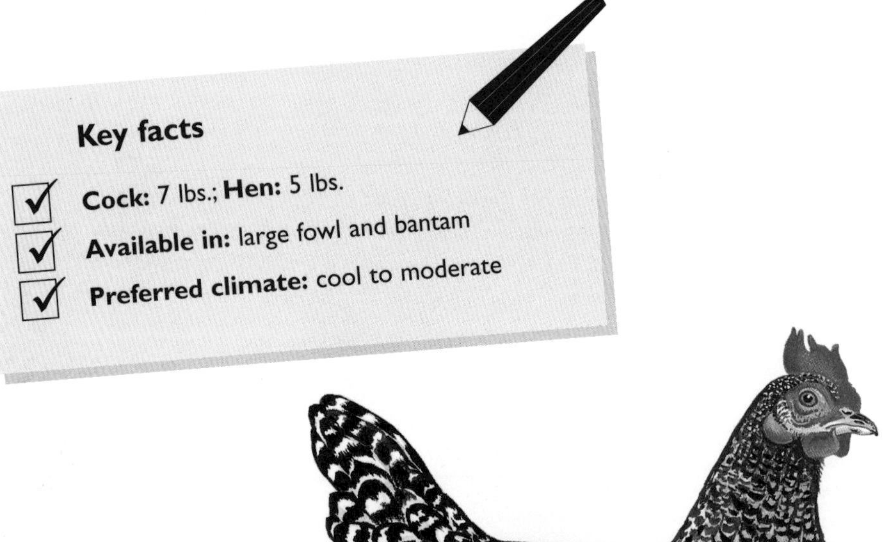

Welsummer

Key facts

- **Cock:** 7 lbs.; **Hen:** 6 lbs.
- **Available in:** large fowl and bantam
- **Preferred climate:** cool to moderate

This attractive bird was developed in Holland. It is a light breed that lays eggs ranging in color from dark brown in some strains to flowerpot red, the color of egg the bird was originally known for. For a light breed it is quite docile and seems to tolerate all types of conditions, including damp, heavy ground. It comes in two colors, partridge being the main one, although some efforts are made intermittently to establish a silver duckwing variety. The partridge has the male with a bright orange-red top and black with brown mottling breast and thighs. The female has a reddish-brown ground color and each feather has small black stippling and a light yellow center shaft. Set off with a bright red single comb and wattles and bright yellow legs it is a breed that is well worth considering.

Araucana

Key facts

- **Cock:** 6 lbs.; **Hen:** 5 lbs.
- **Available in:** large fowl and bantam
- **Preferred climate:** cool to moderate

The reason these rather ordinary looking birds create so much interest is that their eggs are blue to green in color. It may be unfair to call this breed plain because it has a small crest around the rear of its comb, and a small beard and muffling. There are many colors in the breed, the most popular being black, white, and lavender. The latter color is known in some countries as pearl-gray, which could be a more apt description. The female is a light gray but the male has a silver-pearl finish to his hackle. It is a breed that lays plenty of eggs and the males are so active that fertility does not seem a problem. Because the egg color is dominant, crossbred birds are sometimes sold off as real Araucanas. The breed has also been used to create some of the modern blue egg-laying hybrids. There is also a rumpless (tailless) version but because it carries a lethal gene chicks will not always hatch, so this variety is not one for the novice.

Appenzeller Spitzhauben

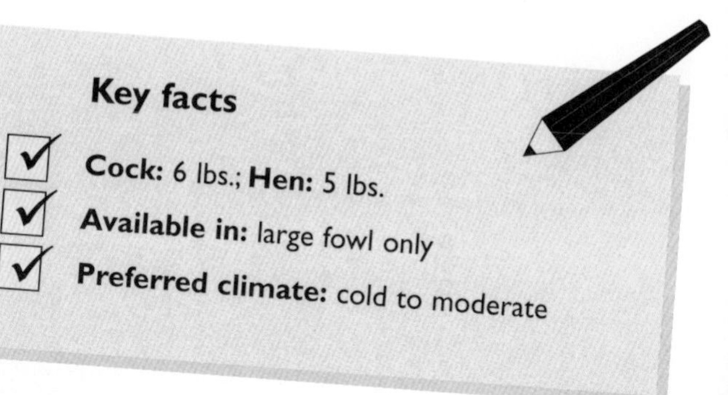

This breed has its origins in Switzerland and has a distinctive appearance in that the top of its head has a narrow upright crest of feathers that point slightly forward, not dissimilar to the headdress of a Roman soldier. Its comb has two upright horns on the front of the head, those of the male being much more visible than those of its female counterpart. It is a rather flighty breed but will reward you with a good supply of medium-sized white eggs. The most popular color is the silver spangled but unlike the Hamburg the spangles are much smaller. There are also gold-spangled, black, and blue varieties. It has blue legs and white earlobes.

Silkie

Key facts

- **Cock:** 3 lbs.; **Hen:** 2 lbs.
- **Available in:** large fowl and bantam
- **Preferred climate:** moderate

This breed is distinguished by its feathers and skin. The feathers are not webbed together as in ordinary chickens but have the same consistency as the fluff around a hen's rear end. The softer the feathers the better, and despite the dainty look of this bird it will not fall ill if it gets a little wet, although shelter should always be available. An even softer version, which has a beard, is also now gaining popularity. This variety is so fluffy it is sometimes difficult to tell its head from its tail! The most popular color is white followed by black, blue, partridge, and gold. The breed has an unusual dark purple skin color that also shows on the comb and face. The earlobes are turquoise. It has five toes, feathers on the outside of the legs, and a small crest that is larger on the female. As well as looking different, it has a reputation for being the most reliable of broodies and its fairly small size means that it will be good with its chicks. It lays a number of batches of small round cream-colored eggs. If it is eggs you're after, then this is not the breed for you but its other attributes can make it appealing, especially to children.

Sumatra

Key facts

Cock: 5 lbs.; **Hen:** 4 lbs.

Available in: large fowl and bantam

Preferred climate: most

While on the subject of dark combs, the Sumatra should be discussed. From the island of Sumatra, this bird usually comes in black, although creamy white and blue versions have also been developed. Besides its dark face it has other distinguishing features, such as a very long tail, often touching the ground in the male, and the legs often have three spurs in a bunch rather than a single one as in other breeds. Its shiny long black plumage makes it a graceful bird and as it loves to fly it looks really good in a large aviary where, if perches are set up high enough, it will happily roost at night. It lays a good number of creamy white eggs that, although fairly small, have a lovely flavor. It, like the Silkie, is a reliable sitter and mother. It is not suitable for small runs as its long tail can get damaged, spoiling its appearance.

Sultan

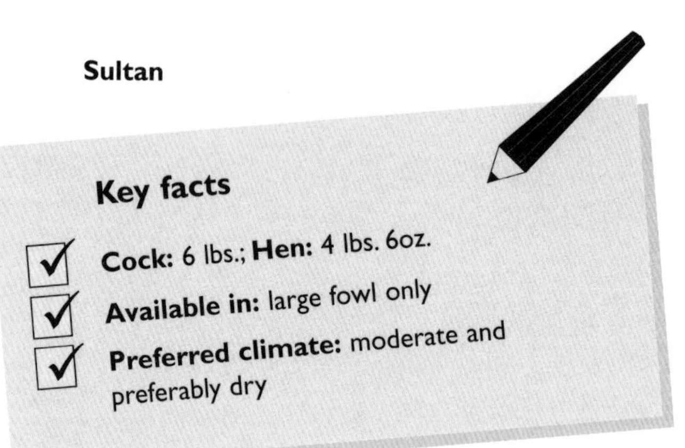

"The breed that has everything" is how you could describe the Sultan. All the adornments or extras that have been described in other breeds are brought together here. It has a crest with beard and muffling and a small comb with two little spikes. The white or pale blue legs and all five toes are covered in feathers. Because the feathers at the rear of the hocks stand out stiffly, these are called vulture hocks. With all these extras it might be thought of as purely ornamental but those who have kept the breed would disagree, being satisfied with its production of white eggs. There is only one color of feather and that is pure white. It is fairly quiet and, although rare, could be worth investigating if you're looking for something different.

Poland (Polish)

This breed is not really for the novice, unless you have plenty of time on your hands. Although it has been around for centuries the breeder's quest for a larger, more rounded crest means that a regime of constant care is needed. The crest must not get wet or dirty as this can cause eye infections or an infestation of mites. Keep the crest clean and treat it with flea powder every two weeks. It flies well so a covered run is best, although many keepers have it under cover. They come in many colors and also in a Frizzle variety. They are not known for producing many eggs.

Key facts

Cock: 6 lbs. 10 oz.; **Hen:** 5 lbs.

Available in: large fowl and bantam

Preferred climate: moderate and preferably dry

Yokohama

This breed is similar to the Sumatra but has an even longer tail, with the male's main sickles dragging along the ground. For this reason it needs to be kept in clean conditions with clean grass or a fresh littered house, although it likes to perch up high, keeping out of any dirt. It does look very attractive, and as the name suggests, comes from Japan. Colors of the Yokohama are white, red, saddled, black red, and duckwing. It has either a single, pea, or walnut comb, the latter being small, curved, and nearly flat to the head. If you are looking for a rare bird that will make observers gasp, then this could be for you. It lays a number of small creamy-colored eggs.

Key facts

- **Cock:** 5 lbs.; **Hen:** 4 lbs.
- **Available in:** large fowl and bantam
- **Preferred climate:** cool to moderate

Onagadori

Key facts

- ✓ **Cock:** 5 lbs.; **Hen:** 4 lbs.
- ✓ **Available in:** large fowl only
- ✓ **Preferred climate:** warm and dry

This Onagadori is kept by only a handful of breeders in Japan although it has received a preservation status from the Japanese government. There are very few specimens in other countries and it demands almost individual care and attention from its owner to preserve the tail feathers, which can be 9 ft. 10 in. in length or longer. The tails on some of the best specimens have been known to reach 32 ft. 10 in. These tremendous lengths are not achieved in one year alone as the bird never molts its tail, unlike all other breeds. It only lays about twelve eggs per year.

Other light breeds

Lakenvelders and **Vorwerks** are like a light breed version of the Light Sussex in silver and gold but with dark slate legs.
Cock: 6 lbs.; **Hen:** 4 lbs. 6 oz.

Marsh Daisies are a British creation made to withstand wet conditions.
Cock: 6 lbs. 10 oz.; **Hen:** 5.5 lbs.

Friesians are an old Dutch breed, light and flighty but good layers.
Cock: 5 lbs.; **Hen:** 4 lbs.

Fayoumis are similar in nature and utility qualities to the Friesians. Created in Egypt, they are reputed to be very disease-resistant.
Cock: 5 lbs.; **Hen:** 4 lbs.

Sulmtaler (right) is an Austrian breed with a small crest and is reputed to be very tame.
Cock: 6 lbs.; **Hen:** 5 lbs.

Bantam breeds

Bantams are small versions of large fowl and must be included when considering chickens to keep because they are so popular and in many cases are found to be equal or better layers than larger fowl.

Miniatures

These are breeds that have large fowl counterparts and come in most of the popular breeds already mentioned, although not all of those classified as rare have bantam equivalents. The definition of a miniature is that it is approximately a quarter of the size of the large. It should have all the same characteristics and probably the same color varieties as their larger counterparts. A bantam male will have a much more shrill crow than the large, which is something to bear in mind, especially if you have neighbors nearby. The bantam hen will lay eggs that are bigger in size respectively than the large fowl. Large fowl eggs weigh between 1¾ –2½ oz. while bantam eggs quite often approach the 1½ oz. mark. You can see here the economics of keeping bantams in that both space and food requirements are considerably less.

True bantams

These are called true because they do not have a large counterpart and are much more ornamental than utility, mostly laying smaller eggs than the miniatures. For those who aren't too bothered about egg production these can fit the bill as a good starting point. True bantams make great pets and are very popular with young children. It is worth noting, though, that the males will often have an even shriller crow than the miniatures!

Dutch

These bantams were originally known as Old Dutch and are one of the smallest trues. They are a very smart little bantam, with the male having a big breast and a very full curved tail that, on top specimens, seems almost half the size of the body. They come in many colors, the most popular being the gold and silver partridge. The males weigh 1–1 lb. 3 oz. and the females 14–16 oz. so you can see just how small a true bantam is. With red comb and wattles, white earlobes, and slate blue legs, they are a good-looking little bird.

Belgian

Belgian bantams come in three types mainly: the d'Uccle, which has a single comb, beard, whiskers, and feathered legs; d'Anvers, which has a rose comb, beard, whiskers, and clean legs; and d'Watermael, which only differs from the d'Anvers in that it has three leaders from the back of its comb. They come in a myriad of colors and vary in weights, with the d'Uccle male weighing around 2 lbs. and the d'Watermael 1.5 lbs. They are jaunty little bantams with a low wing carriage, short back and high tail.

Sebright

This is an English breed with a tradition going back hundreds of years. It comes in two varieties: silver and gold. It is precisely marked with black around the edge of each feather. Besides its markings it is distinctive in being "henny feathered"; in this the male does not have the usual sickle feathers in its tail and any sign of these is seen as wrong. It is a jaunty bird and has a long lifespan.

Rosecomb

This breed has somewhat of the outline of the Dutch bantam but has a rose comb. The most common color is black with rarer blue and white varieties. Other colors are being developed. Besides the neatly worked comb, it has large white earlobes.

Japanese

This breed is very distinctive in shape. It has a large single comb, short back, high upright tail, which is very full, and exceedingly sort legs. The tail in the male is so full it comes forward, almost touching the back of the comb. Some extreme examples come even further forward but are frowned on as "squirrel tailed." The female's tail is also very tall and full but should be upright. It is really just an ornamental breed and should be kept undercover or if outside on short grass. There are many colors, the most popular being the black tailed white, birchen, black, and white.

Nankin

The Nankin is of Asian origin, probably India, and is one of the original bantam breeds that was introduced into England. It is said it was used to create many buff colored varieties. It comes in just the one color, buff, but has both single and rose combs, although the former is the most usual. It is a tame little bird that lays a fair number of white eggs.

Pekin

Probably one of the most popular true bantams, this is seen in some countries in a similar form as a miniature Cochin. It differs from a Cochin in that it has a much more forward sloping carriage with the breast almost touching the ground, especially in the females. It is single combed with feathered legs and long foot feathers spreading out from the sides, especially in the male. They are favorites with children in that they do not fly very much and are tame. They also feel very soft when handled. They lay quite a number of eggs, varying in color from white to tinted. They spend quite a time broody so are reliable in that department. They come in many colors, such as black, white, blue, buff, lavender, partridge, cuckoo, Colombian, and quite a few others. The Pekin should be one of the easiest of true bantams to locate if you want to purchase one.

This, then, is the list of chicken breeds in large and bantams. There may be others that are available in other countries and with lines of communication improving all the time it is not to say that someone won't import a breed that will become very popular. One such breed that is causing much interest and controversy is the Serama, but until the dust settles over the great debate we will just acknowledge its existence. The Serama differs from other breeds in that it has various different weight categories and is said to come in around two thousand colors, none of which are described in the breed's standard. But before we leave the world of hens, let's look at hybrids.

Hybrids

Hybrids are created for the purpose of producing the maximum number of eggs per bird. They are produced in a very selective way by big multinational companies who have invested millions in research and scientific experiments. As a hobbyist poultry keeper, you will not be able to breed anything similar or better by mating one of your own pure breeds with what is thought to be a commercial hybrid male.

Breeding hybrids requires a large number of birds for selection, and the nature of the inbreeding process creates a number of harmful genes in a particular strain. This strain is then crossed with another strain created with the same objective. The bird's genetic makeup means that it has a number of both good and harmful genes. The good genes usually cover up the harmful ones so that two separate made-up strains can combine the best of each to create a hybrid. The bird thus created is the one that finds its way onto the market, but many thousands of adult birds are being kept, tested, and reared to continue or improve upon egg-laying birds. The breeding program is so complex and expensive that the work is highly secretive and so you are only aware of the end result without knowing how it was achieved.

For many years these hybrids were allocated numbers or letters to identify them but now with more new poultry keepers to supply and a desire to personalize the birds, names are used instead so that Speckledy, Black Rocks, or similar are on the list. It was believed that crossing a pure bred male with a group of hybrids would produce beneficial results but this can only happen with some success if a recorded breeding program then continues. If it doesn't, it may be that one generation may produce good results but in the future nothing would be guaranteed unless a program of genetic selection is followed.

A recommendation if you are buying hybrids is to find a registered agent. The problem with buying from someone who has a few look-a-likes is that they may not have the background to produce guaranteed results. If you ask your supplier where they get their birds from as day-olds or growers they should be able to tell you. The producer would then verify this. This warning is given because demand can often exceed availability, giving unscrupulous dealers a chance to cash in.

So hybrids, if they are the genuine article, will be the bird of choice for those looking to maximize the number of eggs from the minimum amount of feed. The performance will almost always exceed that of pure-bred birds. It is worth remembering, however, that hybrids were produced by big corporations quite often not originally involved in poultry but in producing better grain crops that then produced the feed for pure-bred, egg-laying birds. These corporations then wanted to make it an all-encompassing business and so used their scientific facilities to produce hybrids. The hybrid did start with the pure breeds, and with many new hybrids coming on the market, pure breeds still play a part in their production.

Most hybrid producers are listed on Web sites and so can easily be found by those who have access to a computer. You may, however, prefer to browse through poultry or smallholder magazines which have lists of suppliers and often a Web site, too.

Egg production table

The figures below show anticipated egg production for both heavy and light breeds under optimum conditions over a twelve-month period. These figures may not be achieved in a number of breeds if they often go broody.

HEAVY BREEDS	EGGS
Australorp	200
Barnevelder	200
Brahma	150
Cochin	10
Croad Langshan	180
Dorking	150
Faverolles	200
Frizzle	160
Houdan	200
Indian Game	80
Ixworth	100
Jersey Giant	180
Malay	80
Marans	180
New Hampshire Red	220
Orloff	200
Orpington	180
Plymouth Rock	200
Rhode Island Red	220
Sussex	220
Wyandotte	200

LIGHT BREEDS	EGGS
Ancona	240
Andulusian	180
Appenzeller	220
Araucana	200
Fayoumi	260
Friesian	240
Hamburg	240
Lakenvelder	180
Leghorn	240
Marsh Daisy	180
Minorca	240
Old English Pheasant Fowl	200
Poland (Polish)	100
Redcap	240
Scots Dumpy	150
Scots Gray	200
Silkie	100
Spanish	200
Sulmtaler	180
Sultan	180
Sumatra	150
Vorwerk	200
Welsummer	240
Yokohama	150

Q How old is a chicken when it first starts to lay?
A Depending on the breed, you can expect your first egg between the eighteenth and twenty-second weeks.

Q How many eggs, on average, does a hen lay a week?
A Expect a maximum of five to six eggs a week during the first year. This reduces in the second year to around five and further still in following years as the birds tend to lay for a shorter period. Some breeds are prone to broodiness and if allowed to continue sitting, the number of eggs laid greatly reduces. Hens molt around autumn, and during this period they will not produce any eggs.

Q How many years will a hen lay eggs for?
A A hen will lay eggs during most of her life but the numbers will be reduced to an uneconomic level in later years.

Q How long do chickens live for?
A Pure breeds tend to live longer than hybrids and have been known to live for up to thirteen years. Hybrids will not usually make it to more than five years.

Q Can a hen produce chicks without a cockerel?
A Cockerels are necessary to fertilize the eggs that hens produce. A hen will, however, lay eggs without the need of a cockerel—they will just not be fertilized and therefore not turn into chicks.

Q How big should my hen house and run be?
A If looking at the usual small unit for three to six hens, then a 3 ft.-square house with a 5 ft. 11 in. x 3 ft. run is sufficient if moved on a regular basis.

Q Can chickens coexist alongside other household pets?
A Yes. This is possible but is only satisfactory if the other household pets, such as dogs for instance, are kept under control, as a chicken will not be able to stand up to an attack from a much larger animal.

Q Are chickens noisy?
A Unless you are also keeping a cockerel, then the noise levels from hens is not obtrusive. They will only "shout" after laying an egg or if alarmed.

Q Is it unfair to keep just one chicken? Will it get lonely?
A In the majority of cases it is better to keep more than one chicken as it will otherwise wish to spend all its time with you.

Q How do I clean dirty eggs?
A Unless you intend to keep eggs for any period of time you can wash them but do not use hot water. If you want to keep eggs for more than two or three days, then it is best to remove the dirt with a hard brush.

Q How long do eggs keep fresh for and how should I store them?
A Eggs stay fresh for seven to ten days and should be stored in a cool, dark place, or in the refrigerator.

Q Do chickens need their wings clipped so that they don't fly away?
A In most cases if you keep your chickens in a small enclosed run, then it is not necessary to clip their wings. If they are in a large open run and are of light breeds or hybrids it may be necessary to clip the wings to prevent them flying out into danger or damaging your garden. Clipping should only be done on the flight feathers of one wing (to unbalance the bird so that it cannot fly in a straight line) and when the wing is open it should only be necessary to cut through the outer four feathers, about 1 in. from the flesh of the wing. A large sharp pair of scissors is the best tool to use.

Q How can you tell a male from a female chick?
A When chicks start to become fully feathered you will see the feathers on the back just in front of the tail start to become fully formed. A male bird's feathers have pointed ends, while those of the female are rounded. In some breeds, especially the lighter varieties, the male bird's comb will also develop more at this time.

Albumen: also known as egg white. It is the clear part that surrounds the yolk and turns white when cooked.

Anti-coccidiostat (ACS): a food additive that protects chicks against Coccidiosis.

Bantam: a miniature chicken. Bantams are usually about a third of the size of standard chickens.

Broiler: a young, tender male chicken under twelve weeks old, destined to be eaten early on in life.

Broody: a hen that sits on eggs to keep them warm until they hatch.

Coccidiosis: a parasitic protozoal infestation, usually occurring in damp, unclean housing conditions.

Cock: a male chicken after its first molt (usually one year and older). Also known as a rooster.

Cockerel: a young male chicken under one year old.

Comb: the soft, fleshy red growth on the top of a chicken's head.

Crop: a pouch at the base of a chicken's neck where it stores food before it goes into the stomach.

Crop impaction: this happens when feed is stuck in the crop and the chicken cannot swallow to pass the feed down. If the crop feels hard, it could be impacted.

Drinker: a plastic or galvanized container that stores water for chickens to drink from.

Dropping board: a collection unit placed underneath perches in the hen house used to collect droppings for easy disposal.

Feed trough: a shallow container that holds chicken feed. Also known as a hopper.

Gizzard: an organ that holds grit for grinding grain and plant fiber.

Grit: consists of sand and crushed pebbles that chickens are fed to aid the grinding process. Grit is stored in the gizzard.

Hackles: the feathers around a chicken's neck.

Hen house: a house in which chickens live, also called a chicken coop.

Hybrid: a chicken that has been genetically bred from two breeds to acquire the best characteristics, such as prolific laying and meaty carcass.

Layer: a hen in the process of laying eggs.

Laying cycle: the period between the start of lay and molt.

Litter: straw, hay, wood shavings, and shredded paper are all suitable to use as litter. Litter can be used to scatter on the floor of the house or run to absorb any moisture and can also be used as padding for nest boxes.

Molt: chickens molt once a year and during the molting period (which lasts about eight weeks), they shed their feathers and grow new ones. Hens will not lay eggs during this time.

Nest box: a wooden or plastic box designed to encourage hens to lay eggs in it.

Perch: a branch or man-made object on which a chicken goes to sleep.

Pullet: a young female chicken under one year old that has not yet started to lay.

Point of lay: about to start laying, between the ages of eighteen to twenty-two weeks. Some hens will start laying later than this.

Run: an enclosure, often attached to the hen house, in which chickens can safely roam.

Shank: the part of a chicken's leg between the claw and the first joint.

Spur: a sharp pointed protrusion on the back of a rooster's shanks.

Sexing: the process by which the sex of a chick is determined.

Vent: the opening at the rear of a chicken through which all waste matter and eggs pass.

Resources

To find further information about keeping chickens and for suppliers, here are a few online resources:

www.poultryconnection.com/hatchery.htm

www. flemingoutdoors.com/poultry.html

www.chickenkeeping.com

http://smallfarm.about.com/od/farmanimals/a/htkeepchickens.htm

www.ecplaza.net/search/0s1nf20sell/chicken_suppliers.html

To my parents for encouraging me as a young child; my many friends at home and overseas who have shared my enjoyment in this hobby; and those who have encouraged me to write this, my first book. Finally for my wife who has done all my typing and provided me my spare time in this book's production.